Solar Heating and Cooling Systems
Design for Australian Conditions

Solar Heating and Cooling Systems Design for Australian Conditions

E. Baker, C.J. Floro,
J.P. Gostelow and J.J. McCaffrey

School of Mechanical Engineering
The New South Wales Institute of Technology

PERGAMON PRESS
SYDNEY • OXFORD • NEW YORK • PARIS • TORONTO • FRANKFURT

Pergamon Press (Australia) Pty Ltd,
19a Boundary Street, Rushcutters Bay, N.S.W. 2011, Australia.

Pergamon Press Ltd,
Headington Hill Hall, Oxford OX3 0BW, England.

Pergamon Press Inc.,
Maxwell House, Fairview Park, Elmsford, N.Y. 10523, U.S.A.

Pergamon Press Canada Ltd,
Suite 104, 150 Consumers Road, Willowdale, Ontario M2J 1P9, Canada.

Pergamon Press GmbH,
6242 Kronberg-Taunus, Hammerweg 6, Postfach 1305, Federal Republic of Germany.

Pergamon Press SARL,
24 rue des Ecoles, 75240 Paris, Cedex 05, France.

First published 1984

Copyright © 1984, E. Baker, C.J. Floro, J.P. Gostelow, J.J. McCaffrey

Cover design by Pam Brewster
Printed in Australia by Macarthur Press Pty Ltd, Parramatta

National Library of Australia Cataloguing in Publication Data:

Solar heating and cooling systems.

Includes bibliographical references and index.
ISBN 0 08 029852 4.

1. Solar heating — Australia — Equipment and supplies —
Design and construction. 2. Solar air conditioning —
Australia — Equipment and supplies — Design and
construction. I. Gostelow, J.P.

697

Contents

Preface

This book is concerned with the intelligent use of solar energy in practical applications. The emphasis is on materials selection, heat transfer and economics of both individual solar components and complete systems for low temperature applications. Adequate background information is also provided for the system specification, design and installation phases of a solar project.

This text differs from others that are available in two ways. First, it was written specifically for the conditions that exist in Australia and New Zealand. Although excellent books exist for conditions in the U.S. and Europe, they usually contain very few data that are relevant to the southern hemisphere. In this book, correlations that are specific to the northern hemisphere have been modified or replaced with ones that are appropriate for conditions here.

The second way that this book differs from others is that it recognises that students and serious solar designers normally have access to a personal computer. Computer programs (written in BASIC) are given throughout the text which will eliminate much of the tedium that usually accompanies solar calculations. With these programs the lecturer and the students can optimise practical designs using real data. The students can also perform parametric studies on materials and components or evaluate the performance of different types of systems in a practical application. The practitioner can use the programs to help him evaluate the effect of factors that are specific to his site, such as orientation, shading, etc. He can use these programs for preliminary calculations to compare the different systems and components that are available. When a system has been selected, he can use these same programs to optimise and fine tune that system. The programs listed in the text are simplified versions of some of the Apple© programs used on the the SOLARAUST disk available from Pergamon Press.

This book has been used as the set text in a solar course for general engineering undergraduates and as the text for an advanced undergraduate and graduate level solar design subject. It was also used in extension courses run for practitioners, such as architects, builders and manufacturers. To accomodate the varied backgrounds of these three groups, three different analytical techniques are used to solve each type of problem. The three techniques are,

1.   Graphical techniques that require little mathematical background. These techniques are useful for preliminary analyses of solar designs and appeal to the general practitioners with little mathematical training.

2.   Mathematical techniques are developed that can be done on a calculator. These methods give a better grounding in the fundamentals of solar design, and are at a level suitable for engineering and science students.

3.   Finally, techniques developed for the personal computer extend these analytical methods and allow both advanced students and the serious practitioner to perform sophisticated analyses of either single components or entire systems. These programs allow the lecturer to tailor his course to the abilities and needs of his students.

The chapters are organised in a natural sequence. A historical introduction and overview of solar energy utilisation are given in Chapter 1. Materials selection, system reliability and passive design are also included in the first chapter. In Chapter 2 the methodology for heating load calculations is introduced. A mastery of this material is considered essential for any realistic attempt to analyse or design a solar thermal system. The application of calculation methods can only be as good as the assumptions and data base employed; the purpose of Chapter 3 is to provide needed information on the thermal environment. The one essential component for all solar energy systems is the solar collector. Some information on different types of collectors and on improvements in design and materials is presented in Chapter 4. A basis for evaluating the type and size of collector appropriate for the various applications is given in Chapter 5. The question of storage of collected solar energy is a decisive one for the effectiveness of most systems and is addressed in Chapter 6. The use of solar air conditioning may be important for Australian conditions and is the subject of Chapter 7. The specific equipment required for this purpose is considered in Chapter 8. Ancillary equipment used to distribute and control the working fluids is considered in Chapter 9. Also considered are the questions of freeze and overheat protection. The performance of a system will be influenced greatly by the original specification and installation and these are the topic of Chapter 10. Finally comes the all-important bottom line. Regardless of the technical merit of a system it must be economically viable. Procedures for costing and for performance analysis are presented in Chapter 11.

This book has been a team effort with responsibility and credit for the content of the various chapters being allocated as follows: Earl Baker was responsible for Chapters 1, 4, 6 and 11; Cesar Floro for Chapters 2, 3, 8 (part) and 10 (part); Paul Gostelow for Chapters 5, 9 and 10 (part) and John McCaffrey for Chapters 7, 8 (part) and 10 (part). Earl Baker was also the editor and he wrote and developed all of the computer programs listed in the book. Preparation of the book involved the support and co-operation of many people. The drawings were done by Charles Evans and Lex Metcalf and the bulk of the typing was done by Beth Cook. In addition, suggestions by students that took the course have significantly improved the book over earlier versions.

E. Baker

C.J. Floro

J.P. Gostelow

J.J. McCaffrey

Sydney, September 1983

CHAPTER 1

Solar Energy Utilisation
Introduction and Overview

AIM

The design and optimisation of efficient solar heating and cooling systems is a complicated process. Although the basic principles of solar design are relatively simple, their successful execution requires a thorough understanding of the application of these principles to practical systems. The primary aim of this book is to provide the basic information needed to buy or select solar heating and/or cooling systems. The selection and assembly of solar equipment into complete systems together with the design optimisation of those systems will be emphasised. All of the information needed to make an economic analysis of a solar heating or cooling system and to predict its basic operating characteristics is contained in this book. The main emphasis will be on domestic systems; however, the same principles apply to most industrial systems.

HISTORICAL BACKGROUND

It has often been said that if man does not learn from history, he will be doomed to repeat many of those mistakes. During its long history, the U.S. solar industry has made many costly mistakes. The history of the solar domestic hot water industry in the U.S.A. will be reviewed to help us avoid repeating these mistakes.

Solar domestic hot-water systems were introduced in the U.S. at the end of the last century and since that time considerable design experience has been accumulated. Unfortunately most of this experience has been gained through design or marketing mistakes which gave solar water heaters the reputation of being uneconomic and unreliable [1]. Between 1891 and 1930, solar water heaters were widely used in California. The system used then was quite similar to the flat plate collector system of today. However, during the 1920's, the discovery of local natural gas priced solar heating systems out of the California market. The solar technology developed in California was was also used in Miami, Florida. In the depression of the 1930's, many solar manufacturers were forced to reduce their costs. For example, some companies reduced costs by spot soldering the water tubes to the copper sheet of the flat-plate collector and this gave inadequate heat conduction. Many contractors reduced costs by laying the collectors directly on the roof and, since the seals were not completely watertight, rain accumulated beneath the collectors. Eventually, since the water could not run off or evaporate, the bottom of the box rusted and both the collector and roof had to be replaced. Some suppliers also reduced the size (and price) of the units to a size that was too small to be efficient. All of these factors gave the solar industry such a bad name that in the mid 1930's, the U.S. Federal Housing Authority had to investigate. As a result, national standards were adopted for the manufacture and installation of solar water heaters and guidelines were set for collector size and tank capacity. With these safeguards, the solar water heating industry again thrived in Miami and perhaps as many as 40 000 units were installed between 1935 and 1941. In 1941, solar heater sales in Miami were double those of conventional heater sales, and the solar industry appeared to be well established.

During World War II, the shortage of copper brought solar water heater production to an abrupt halt. After the war, just when the solar industry was ready to move again, solar storage tanks began to burst on systems that had been installed during the late 1930's. It was found that the pressurised hot-water flowing between two dissimilar metals (the copper tubing and the galvanised iron tank) had caused the tank to corrode. Often the tank would burst around 2 a.m. when the mains pressure was at its maximum. If the tank was located over an important part of the house, the water damage could be quite costly. Many of the tanks burst only 8 to 10 years after their installation. Because the majority of the systems were installed in the late 1930's, the plague of tank failures began in the late 40's and lasted into the 1950's. As a consequence of these failures, the solar industry in Florida was virtually destroyed by the late 1950's.

The energy crisis of the 1970's gave new life to Florida's solar industry.   However, during the extremely cold winters of 1977 and 1978 many solar collectors froze and burst, which set the industry back yet again.   Before 1977 it was believed that Miami, like Sydney, would "never" have a hard freeze.   Use of anti-freeze solutions in the collectors and specially lined tanks will probably protect the industry from these catastrophes in the future, but, in the minds of many Floridians, solar heating systems presently have a reputation of being unreliable.

The history of solar air heaters, although much shorter, has also been disappointing.   Even when an air system is designed by an expert, its performance is still often found to degrade significantly with age [2].

SOLAR HEATING SYSTEMS

Liquid and Air Systems

Solar heating systems usually consist of collectors to trap the solar radiation, some fluid to transport the heat from the collectors to where it is needed, a heat storage unit, pumps, blowers, heat exchangers, control systems, etc.   In most homes it is also necessary to install an auxiliary heating system for those times when the solar heating system is not able to carry the entire load.   Air-heating and water-heating systems are both commonly used and, although they both may perform the same overall function, they differ significantly in their design and operating characteristics. A diagram of each of these systems is shown in Figure 1.1.   The overall performance is roughly the same for the two types of systems.

Liquid systems

The main advantages of a liquid system are that it uses a low-cost fluid with a high heat capacity and it only needs small diameter piping (rather than large air ducts).   Also, since solar hot-water

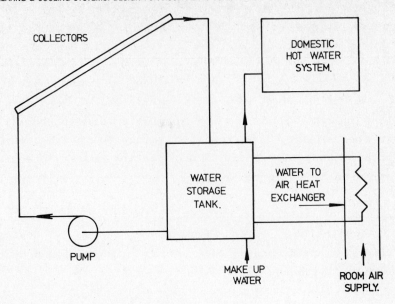

(A)   LIQUID   SYSTEM.

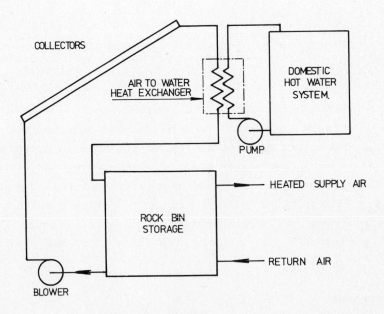

(B) AIR   SYSTEM

Fig. 1.1.   Diagrams of solar energy systems for
water and space heating.

systems have been in use for nearly a century, many different types of liquid systems have been developed and extensively tested, and many of the problems have already been solved.   In addition, liquid systems are compatible with the solar air-conditioning systems presently being developed.   Unfortunately, the most commonly used liquid (water) has certain undesirable characteristics.   Its high freezing point, low boiling point and tendency to corrode metals require designs and materials which increase costs significantly.

Air systems

Air systems avoid the problems of corrosion, freezing, boiling, fluid replacement and potential damage caused by water leaks. However, they do require larger volumes for thermal storage (approximately three times as much as required for water) and air ducts require much more space than water pipes.   The power required to drive the fans is also much greater than is required by the pumps in an equivalent liquid system.

The heat delivered per square metre of collector area is about the same for the air system as for the liquid system, and the total installed cost of air systems appears to be about the same as the cost of liquid systems for the same heat delivered.   However, the value of the extra space and the extra power required by the hot-air system tends to make commercially installed air systems more expensive than the equivalent liquid systems.

THE COLLECTOR

For any domestic systems, the maximisation of the collector efficiency (or of the operating temperature) is not of primary importance.   The main criterion for collector selection is how much heat can be collected for each dollar invested.

There are two basic types of solar collectors, the flat-plate type and concentrating type.   The latter uses reflectors or lenses to concentrate the solar radiation onto an absorber.   These collectors must track the sun, and usually cannot concentrate the diffuse portion of solar radiation that is reflected from the earth and sky. Consequently, they are generally only economical where high temperatures are required and are rarely used in domestic systems.

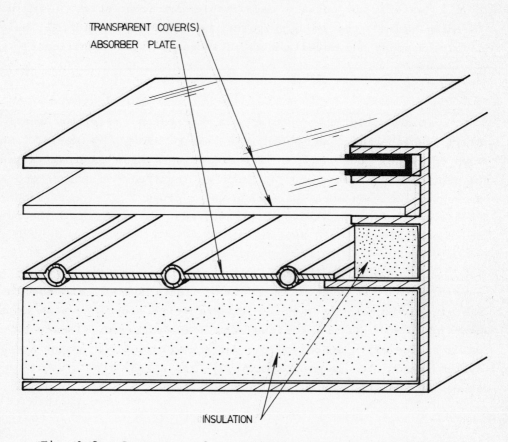

TRANSPARENT COVER(S)
ABSORBER PLATE

INSULATION

Fig. 1.2.   Structure of a water-cooled solar collector.

Solar collectors also differ in the methods used to reduce the heat loss from the collectors.   Since the absorber is heated to a higher temperature than the surroundings, an absorber loses heat by conduction through the insulation beneath the absorber and by convection and radiation from its top surface.   When only a small temperature rise is required (such as with swimming pool heaters) the heat lost from the top of the solar absorber may not be significant.   At higher operating temperatures, one or more layers of glass may be placed over the absorber to reduce the convection losses (see Figure 1.2).   Since the glass also reflects and absorbs some of the incoming solar radiation, the reduction in heat losses must compensate for the reduction in insolation reaching the absorber.   The convection losses can be reduced significantly by evacuating the air from around the absorber, but these evacuated systems are usually too expensive for home heating applications.

THE THERMAL STORAGE SYSTEM

Since the sun doesn't always shine when it is needed, domestic solar heating systems usually have some means of storing thermal energy. Although many possibilities exist, most liquid systems rely on hot water storage and most air systems store heat in rock beds for reasons of simplicity and economy.   There has been considerable work done on phase-change systems that use the large latent-heat-of-fusion of waxes, salts, and other chemical compounds. This latent heat is released when liquids solidify (or is stored when solids melt) and is considerably greater than the heat stored or liberated when an equal volume of water or rocks changes temperature by (say) 50°C without changing phase.   Consequently, phase-change heat-storage units can be significantly smaller than the other types of thermal storage units.   Unfortunately, technical difficulties and economic disadvantages have yet to be overcome before phase-change heat-storage systems become practical for use in domestic applications.

SOLAR COOLING SYSTEMS

INTRODUCTION

The effectiveness of a cooling system is usually expressed in terms of its coefficient of performance (COP).   The COP of a system is defined as the ratio of the heat removed from the space to the energy supplied by external sources.   It represents the ratio of the cooling effect achieved by the system to the energy that is paid for.   The conventional vapour compression system characteristically has a COP of about two, but it can be as high as four.   The most common solar cooling system is the lithium-bromide-water absorption refrigeration system.   This system has a maximum COP of about 0.8, but more typically operates below 0.6.*   Consequently, for these systems, more energy is required to run the system than is removed from the conditioned space.

Either conventional electrically driven vapour compression systems or gas fuelled absorption systems may be converted to use solar energy.   However, at present, only the lithium-bromide-water absorption system is commercially available for residential space cooling applications.

ABSORPTION REFRIGERATION

The absorption system, like the vapour compression system, accomplishes cooling by expansion of a liquid from a high to a lower pressure (and temperature).   Evaporation of the liquid refrigerant absorbs heat from the surroundings.   Absorption systems use inorganic refrigerants (e.g. water or ammonia) together with an absorbent.   An absorbent is a liquid that combines chemically with the refrigerant.   In the lithium-bromide-water system, lithium

---

*The COP of these two systems cannot be compared directly since the "energy paid for" is supplied by electricity in one case and by solar energy in the other.

bromide is the absorbent and water is the refrigerant; whereas in the ammonia-water system, the ammonia is the refrigerant and the water is the absorbent.   A detailed explanation of the operation of absorption systems is given in Chapter 7.

REVERSE-CYCLE AIR CONDITIONERS

Air conditioners absorb heat from inside the building and reject this heat to the outside air, thus cooling the building and heating the outside.   By reversing this operation, the system can be used during winter to heat the building air and cool the outside air.

In the cooling mode, the vapour compression system in Figure 1.3A cools the room air as it is circulated over the evaporator coils. The warm refrigerant vapour from the evaporator is first compressed (to raise its condensation temperature) and then condensed.   During condensation this high-temperature heat is rejected to the outside air.

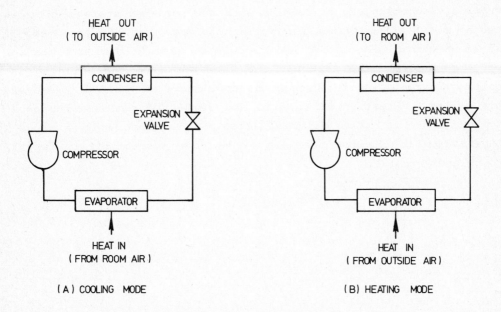

Fig. 1.3.   Vapour-compression reverse-cycle
air-conditioning system.

In the heating mode, Figure 1.3B shows that the coil inside the house becomes the condenser and the outside air becomes the heat source for the evaporator. In this mode, solar energy could be used to increase the heat source temperature and thus reduce the electrical energy required by the system.

EVAPORATIVE COOLING

When water evaporates in air, it absorbs energy from the air and reduces the air temperature. On hot, dry days warm air can be cooled by passing it through an air washer. For example, outside air at 38°C and 20% relative humidity can be cooled in an air washer to 25°C, but the relative humidity would become an uncomfortable 71%. Consequently, evaporative cooling is only effective when the relative humidity is low and the required reduction in temperature is not too large. However, when it can be used, it is an efficient way of cooling since the only power required is that used by the fan.

Evaporative cooling can be coupled with the rock-bed storage unit of a solar air-heating system for summer cooling. In this case, night air is evaporatively cooled and circulated through the rock bed to cool the rocks. The next day the warm air from the building is cooled by circulating it through the cool rocks in the bed. Use of this system is restricted to arid and semi-arid regions with cool nights and low relative humidities. In more humid areas, condensation in the rock beds may make them susceptible to fungus growth and vermin.

DESIGN OF PASSIVE SYSTEMS

INTRODUCTION

Solar space-heating systems contain some type of collector, a means of heat storage and some medium (e.g. water or air) to circulate the heat. They may also contain additional heat exchangers, pumps, blowers, etc.   Passive systems have many of these components integrated directly into the building, but use little mechanical hardware and require little or no electrical energy to operate. The passive building is designed to take maximum advantage of the sun's energy to heat the building in winter and to provide cooling and ventilation in the summer. Air circulation is usually accomplished by natural convection. In the U.S. the owners of some passive homes claim a saving of 80% on their fuel bill, but the average is probably closer to 40%.   In new homes, passive features tend to cost less than active systems for the same total energy delivered, but passive design does restrict the configuration of a house, its position on a particular site and the type and location of trees around the house. Passive control requires frequent adjustment of shading devices, so either expensive control systems are needed or the home-owner must manually adjust the shading devices to suit the availability of sunlight and the room temperature. Passive design techniques are not effective in regions such as Darwin and Cape York where both the temperature and the humidity are high, as these techniques do little to decrease the humidity in a home.

Passive design begins with the choice of site and the orientation of the house.   The location of the house is always a compromise between the need for north facing windows for heating in the winter and the need to capture the cool summer breezes.   Near a large body of water, breezes tend to move from the water to the land during the day and in the opposite direction at night.   On a north-facing hill, breezes tend to blow up the hill during the day and down at night.   For other sites, it may be possible to obtain wind data in the form of "wind roses" from the local weather bureau.   As an example, the summer and winter wind roses for Sydney are shown in Figure 1.4.

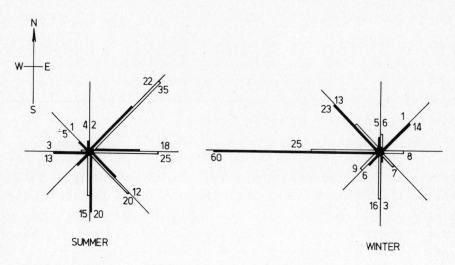

Fig. 1.4. Wind roses for Sydney.

In this figure, the direction of the wind is toward the centre of the rose, the solid bars indicate the morning wind and the open bars the afternoon winds.   The length of the bar shows the proportion of time that the wind is in that direction and the number indicates the per cent of the winds that exceed 32 km/hr.   The figure shows that the wind is from the west or northwest during a large proportion of the winter so it is worthwhile to have a wind break to the west of the house during the winter months in Sydney.   Further, during the summer months, it is desirable to block-out the hot afternoon sun which is also from the west, so a permanent screen on the west side of a house such as a wall of evergreen trees would have a significant effect on both the heating and the cooling bill for a typical Sydney home.   However, since the west side of one home is usually the east side of another home, the same wind screen that blocks the cold westerly winter winds from one home may also block the cooling easterly summer breezes from the neighbouring home.

In addition to careful orientation, well designed passive structures are also well insulated with windows and awnings appropriately

placed.   Materials that absorb and store solar energy are used
extensively and the overall design may incorporate earth berming,
deciduous trees and other techniques to control the heat gained or
lost by the structure.

INSULATION

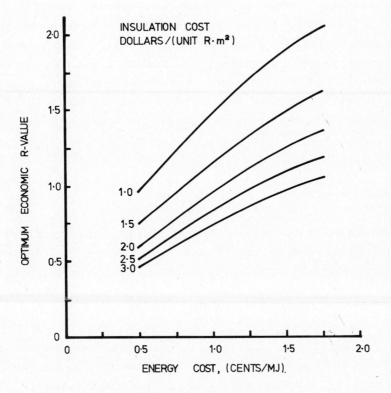

Fig. 1.5. Optimum economic R-value for ceiling
insulation in Sydney.

When energy was cheap, energy conservation in structures was usually
ignored. Now that energy is expensive, the addition of insulation,
particularly to ceilings, will usually significantly reduce heating
costs, and reduce temperature gradients within the home.   Glass
fibre, rock wool and cellulose fibre are the most commonly used
insulations. Glass fibre and rock wool are available as batts or as
loose fill.   Both batts and loose fill can be used on horizontal
surfaces, but loose fill tends to settle with time so it should not

be used in vertical spaces (e.g. wall cavities). Cellulose fibre is only available as loose fill and is generally pneumatically applied onto ceilings.

O'Brien [3] of the CSIRO performed a present-worth analysis to determine the optimum economic ceiling insulation for Sydney in terms of the R value. In this analysis, O'Brien assumed that the repayment period and effective life of the insulation were 20 and 40 years, respectively. The fuel cost was assumed to rise at an annual rate of 1% above inflation, the discount rate used was 10% and the repayment interest rate was taken as 12%. For Sydney, the optimum economic values of thermal resistance are shown in Figure 1.5 as a function of initial fuel cost and insulation cost [3]. The figure indicates that at the current cost of electricity, the optimum R value for ceilings in Sydney is between 1 and 2 (m^2 $^{\circ}$C/W).

TYPES OF PASSIVE SYSTEMS

Modern construction practices tend to minimise the mass of the structure and thus reduce its "thermal inertia" or its ability to store heat. Compare, for example, a modern brick-veneer home with some of the old sandstone structures (or even a double-brick structure). The sandstone structure tends to maintain a more uniform room temperature during the daily variations in outside temperature, mainly due to the heat storage capacity of its massive walls and floor. This concept of a large thermal mass is a primary feature in most of the current types of passive systems. Other design features such as small windows, high ceilings and good cross-flow ventilation help to keep the sandstone home cool in the summer, but some of these features also tend to waste heat in the winter.

Direct Gain Systems

Direct gain passive systems (shown in Figure 1.6) are probably the simplest of all the passive systems. They consist of windows that are placed to maximise the glass area that "sees" the winter sun, a

large thermal mass to store the heat and a layer of insulation on
the outside of the thermal mass to keep the heat in the building.
The roof overhang allows in the maximum amount of solar energy
during the cold winter months when the sun angles are low and shades
the windows in the summer when the angles are high. Usually,
thermal shutters or curtains are used on the windows to reduce the
heat losses at night.

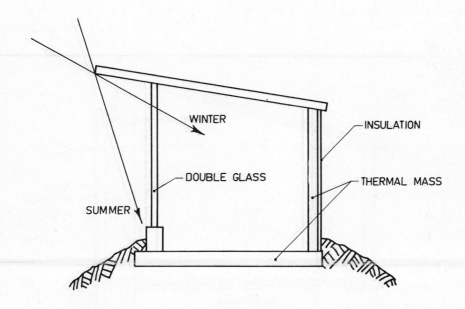

Fig. 1.6.   Direct gain system.

Thermal storage reduces the daily temperature variations of the
structure and provides a means of saving the excess heat received
during the day until the night when it is needed. The absorbed
solar radiation will raise the temperature of a small thermal mass
by several degrees while the same quantity of solar energy will
raise the temperature of a large thermal mass by only a few degrees.
By reducing the magnitude of the daily temperature swings, the
building becomes more comfortable.

As a rough "rule of thumb" 2.5 to 3 square metres of a well designed direct-gain window is equivalent to one square metre of collector area in an active system [4].   These systems require more than 150 kg of water or 700 kg of concrete or rock should be used for each square metre of north facing glass.   If the thermal storage does not receive direct sunlight, up to four times more storage capacity may be required.   Thus, if a slab floor is used for the thermal mass, any furniture used in the room should be placed so that it does not "shade" the floor.   Also, carpets must not be used.

Thermal Storage Wall

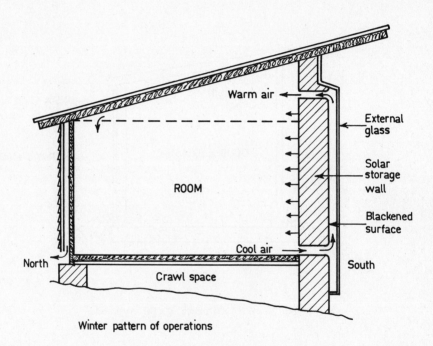

Winter pattern of operations

Fig. 1.7.   Trombe wall.

A second type of passive system which operates by increasing the thermal inertia of the building is the storage wall such as the Trombe Wall shown in Figure 1.7.   With this system, the radiant energy passing through the glass is absorbed by the black concrete surface.   When the wall is heated to a few degrees above the

temperature of the air in the room, the air will flow by natural
convection through the bottom air vent, up between the wall and the
glass, and back to the room through the top vent, gaining heat as it
passes the wall.   The Trombe Wall has many of the shortcomings of
other passive devices, such as inefficient use of storage mass and
wide variations of house temperatures.   It has the added problem
that heat is stored on the side of the wall that is exposed to the
cooler outdoors.

Since water has about twice the heat capacity of an equal volume of
concrete, containers filled with water have been effectively used to
increase the thermal inertia of buildings.   Water walls have been
constructed which use 40 gallon oil drums to increase the thermal
mass of a structure.   However, the construction complexity and cost
of these water filled systems probably make them uneconomic for use
in much of Australia.

Solar Greenhouse

If a greenhouse built on the north wall of a house (such as in
Figure 1.8) incorporates some type of thermal storage wall, it can
provide heat to the house.   For most of the year solar radiation
absorbed in the greenhouse will provide all of the heat required for
the plants and a substantial amount of energy will be available to
help heat the house.

Solar Roof Ponds

In the type of passive solar system shown in Figure 1.9, thermal
storage is accomplished in water-filled bags located in the ceiling
of the building.   Movable insulating panels are used to retain the
absorbed solar radiation during the winter months and to reflect the
solar radiation during the summer months.   The system can also be
uncovered on summer evenings to take advantage of nocturnal
radiation for cooling.   This system has worked very well in a small
building without any backup in the rather mild climate of
Atascadero, California, but the structure had to be reinforced to
support the waterbags on the roof and the house was 20% more
expensive to build than a conventional home.

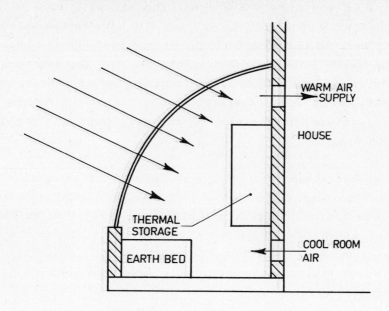

Fig. 1.8 Solar greenhouse.

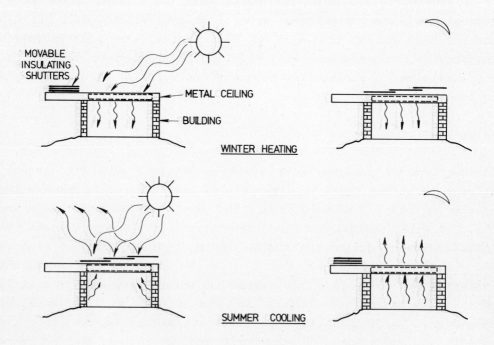

Fig. 1.9. Operating modes for a solar roof pond.

THERMAL CONTROL OF PASSIVE SYSTEMS

The biggest problem in the design of passive solar energy systems is
the storage of heat and the control of the rate of heat release to
maintain suitable comfort conditions. The situation is a paradox in
that storage of sensible heat requires a temperature change within
the storage medium while the object of thermal control is to
maintain a constant temperature. One approach is to use such a
large thermal mass that the variations in temperature are tolerable.
A less favoured approach is to insulate the storage mass from the
living space. With either approach, the normal temperature
variations experienced in passively designed homes are significantly
greater than those normally experienced with active mechanical
systems.

The passive design techniques discussed above may be used to reduce
fuel bills and to make most conventional structures more
comfortable. Often these passive features are chosen for
architectural or personal reasons and the economics of the feature
is not a significant consideration. More often, the passive
feature is expected to pay for itself. In that case, an
experienced architect/designer should be consulted that has a
thorough understanding of the physical principles involved and and
the experience needed to apply these techniques. Even then, most
of these features will prove to be uneconomic for the environment
that exists in the larger Australian cities.

ANALYSIS AND OPTIMISATION OF PASSIVE FEATURES

Although the principles of passive design are quite simple, the
actual design procedure for an efficient, comfortable and economic
passive structure is complex. The most commonly used procedure for
the analysis of a passive design is the Solar Load Ratio (SLR)
method [5]. The SLR is the ratio of the solar heat absorbed in a
building for a given time period (usually a month) to the heat
losses during the same period. From this SLR the Solar Savings
Fraction (SSF) can be calculated. The SSF is the ratio of the
energy saved by the solar feature to the total energy required by an

equivalent building that does not have this passive feature. Thus, the SSF is the proportion that the passive feature contributes toward heating or cooling the building. For fine-tuning a structure this analysis becomes quite involved and is best done on a computer. Computer programs that operate on mainframe computers have been in use for several years and are available at most Australian universities. Recently, a passive design software package called Solarsoft [6] was developed for use on the APPLE© computer. This package offers a quick and inexpensive alternative to mainframe simulation methods. It is quite simple to use and is intended for architects/engineers and owner/builders involved in the design of passive solar structures. It has received good reviews in the U.S. [6] and is suitable for use in Australia. Consequently, it is now possible for the architect/designer to perform his own optimisation analysis for each passive solar feature in a structure and to predict the real cost and savings for the feature. To operate this program the user must supply data on the heat losses and gains of the structure, its shading and location, as well as the climate. Since all of these data are also needed to design an active solar heating/cooling system, this book can be used in conjunction with Solarsoft to predict the performance of passive features for Australia and New Zealand.

PRACTICAL LIMITATIONS ON PASSIVE DESIGN

Passive design can be used in many homes; however, for most homes, energy conservation and passive design cannot provide the whole answer. For example, most of the homes that will exist forty years from now have already been built and, unfortunately, the most cost effective energy conservation and passive solar techniques are usually difficult to apply to existing homes. Site orientation, window size and position, wall insulation, and the use of slab floors can only be adjusted freely before the house is built. In addition, many sites do not lend themselves to a northerly orientation without windows on the western side. Consequently, most homes require some form of active heating system (either conventional or solar).
In those cases where the site has its best view to the north (or has

no view at all), where there are no neighbours, trees or other obstructions to sunlight and where the shape of the site is suitable for passive design, there are still economic and psychological limits to the number and type of passive features most people will accept.   Bright sunlight can cause glare problems and may cause materials to fade and deteriorate.   Also most people tire rather quickly of opening and closing drapes and of shifting scatter rugs to allow the sun to strike the slab floor.   People sometimes feel that the reduced amount of fresh air in passive homes (e.g. CSIRO uses 0.4 air changes per hour in its demonstration house in Victoria [5]) can causes a build up of "indoor air pollution" from cooking, smoking, and from building materials.   In addition, dampness, mildew and airborne germs can be more of a problem when the infiltration rate is kept low.

As a specific example, consider the instructions given to occupants of a low energy demonstration home in Victoria [6].   This house was designed to use only half of the energy consumed in a typical Ballarat home of the same size;   however, the booklet warned that a careless user demanding high comfort conditions could treble the amount of energy consumed.   In this booklet the occupants were advised to:

1. Keep the doors to rooms and closets closed when
    they are not in use.

2. Open and close internal doors, windows and curtains
    often to reduce heat losses in winter and heat gains
    in summer.

3. In summer, keep windows and curtains shut from morning
    until early evening and open doors as little as possible.

4. Remove and replace the shading devices on the pergola
    twice a year.

5. Wear warm clothes indoors during the winter (a $1^{o}C$
    reduction in the thermostat will give a 3% saving in fuel).

6. Bedrooms and bathrooms are not heated, so electric
    blankets and small portable heaters may be required.
    (This will make it more difficult to use these rooms
    as children's play areas, studies, hobby areas, etc.)

7. Do not do more than two loads of wash in any one day and
    then only if the previous day was reasonably sunny.

To have to adjust ones life style to this extent just to save 50% of
the energy bill may be unacceptable to many people.

For these reasons energy conservation and passive design techniques
will probably reduce the energy requirements of the average home
built in the future by no more than 30 to 40% (rather than the 70 to
90% claimed by some people today) and much of this gain will be due
to the proper use of insulation and the reduction of infiltration.

RELIABILITY OF SOLAR HEATING AND COOLING SYSTEMS

EARLY SYSTEM FAILURES

Although solar heating systems have been in use for many decades, design errors still occur which cause many systems to malfunction during their first few years of operation. For example, a study by Argonne National Laboratories [7] of 66 new solar heating and cooling systems for the period between July 1978 and April 1979 found that:

1.   20% of the systems froze.

2.   The controls failed on 28% of the systems.

3.   Interconnection failures occurred on 21% of the systems.

4.   The collectors on 25 of the 66 systems had a total of
      47 different problems that had to be corrected during this
      period.

The fact that these early failures still occur after nearly a century of design experience indicates the need for independent testing of solar components.   It also shows that a buyer should insist on a reasonable warranty before purchasing a solar system.

LONG-TERM FAILURES

After the system has been in use for many years long-term failures will begin to occur. Thermally-induced fractures, ageing, sedimentation and corrosion all fall into this category. Thermally-induced solder-joint failures, such as the one shown in Figure 1.10, may be caused by collector overheating, by use of the wrong solder or by poor soldering techniques. The other types of long-term problems can be minimised by the proper choice of materials and coolants.   Closed-cycle systems may need corrosion inhibitors and other chemicals in their coolants.   Since these chemicals are toxic, they cannot be used in open-cycle, domestic hot-water systems. In such cases, materials of construction and other design features determine the ultimate life of the system. Copper, aluminium, stainless steels, mild steel, brass, plastics and paints have all been used in solar heating and cooling systems.

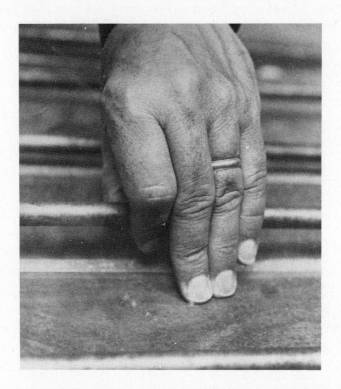

Fig. 1.10. Solder failure in tube-to-plate bond
(courtesy of Argonne National Laboratories).

Copper

Copper, as tubing and plates, is probably the most commonly used material in solar heating systems. Before any metal is used, the designer should ensure that the metal is compatible with the other parts of the system. In systems that contain more than one metal, galvanic corrosion of the less noble metal can occur at the junction between the two metals. Aluminium, steel and galvanised iron are all less noble than copper, and will therefore corrode in preference to the copper. If the two metals can be electrically insulated from each other, galvanic corrosion will not occur.   Unfortunately, most practical electrical insulators (e.g. plastics) deteriorate too rapidly in this type of environment to give the 20 to 30 year life required of solar-heating systems.   Also, many coolants attack copper and the dissolved copper can precipitate on the less noble metal and cause localized corrosion.

Copper may also degrade some of the glycol mixtures used in antifreeze which can lead to bacteriological growth in the solution. Corrosion inhibitors should be used with these solutions and the acidity of the fluid should be closely controlled.

In areas with hard water, scaling or furring may block the copper flow passages prematurely and may adversely affect the heat-transfer characteristics of the collector and of any heat exchanger used in the system.

Aluminium

Many American systems use roll-bonded aluminium sheets for their collector plates.   Use of these plates is usually restricted to closed systems since aluminium is susceptible to pitting corrosion in water, particularly when copper ions are present.   Often de-ionised water with a corrosion inhibitor added is used in these systems and the aluminium is electrically insulated from the rest of the system.   The use of aluminium in direct contact with water in an open system should be avoided.

Aluminium sheet is also used for the collector housing.   In this case, care must be taken to insure that the aluminium is not in contact with other metals or with alkaline building materials.

Stainless Steels

Stainless steels are generally compatible with copper and aluminium. Although stainless steels often resist corrosion, they can be susceptible to crevice corrosion, stress corrosion, pitting and intergranular attack.   Many of these forms of corrosion have occurred when stainless steels were used in solar heating systems,

often as a result of weld sensitisation.   Usually these problems
are a direct result of inadequate understanding of the materials
used.   Attention to design details such as the formation of
crevices, welding, removal of weld spatter, etc., can improve the
life of these systems significantly.

A new family of ferritic stainless steels called "super ferritics"
has superior resistance to pitting and crevice attack and holds
promise for solar heating applications. Some American manufacturers
offer a 10 year guarantee on solar collector panels made with this
alloy.

Steel

Rusting generally makes bare iron and steel pipes unsuitable for
open solar heating systems. In addition, the dissolved rust will
stain bathroom fittings.

Galvanised steels have been used successfully in closed systems
where corrosion inhibitors and corrosion resistant heat exchangers
can be used. Galvanised steels can also be used in open systems
when the pressures and temperatures are not excessive.   At
temperatures above $65^{o}C$ the zinc in the galvanised coating no longer
protects the steel and under some conditions may actually promote
corrosion.   Consequently, the water temperature should generally
not exceed $55^{o}C$ in these systems.   In addition, soft water
accelerates corrosion of galvanised steels.   In soft-water areas,
open systems that use galvanised steel will probably have a very
short life. Also, a collector housing made with galvanised steel
could corrode after a few years unless the housing is painted.   In
general, unpainted galvanized steel should not be used in solar
water heating systems.

Brass

Brass fittings are often used in solar heating systems. Care should
be taken in selecting these fittings as dezincification may occur in
brasses with a zinc content above 30 per cent.   When
dezincification occurs, holes may be formed which can produce leaks

from the system or the zinc corrosion products may collect in the system and obstruct the flow.

Plastics

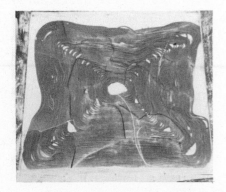

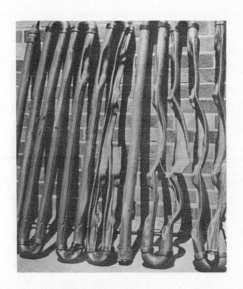

    a)   Creep in plastic   b)   Creep in plastic
           solar air heater.   solar water heater.

    c)   Thermally-induced
         crack in plastic
         collector cover.

Fig. 1.11. Plastic degradation due to high stagnation
temperatures.

Most engineering plastics will not survive for long periods (e.g.

over 10 years) in solar heating systems.   After a few years in this
type   environment,   engineering   plastics   often   experience
photodegradation, stress corrosion, embrittlement, and excessive
stress relaxation or creep.   Figures 1.11(a) and (b) show the
effect that the high stagnation temperatures attained within glazed
solar collectors can have on plastic tubes and pipes.   Since the
thermal expansion of plastics is usually much greater than that of
metals, thermally induced stresses can also cause system failures,
such as the fractured acrylic cover shown in Figure 1.11(c).   In
addition, the ageing characteristics of plastics vary significantly
in different environments, for example, acrylics that survived 17
years of tests in the desert, failed after a few months of use in
Southern   Florida.   Consequently,   the   use   of   plastics   in
solar-heating applications should be viewed with caution unless
long-term tests conducted on that plastic show that it can safely be
used in that specific environment.

Paints

Fig. 1.12.   Example of paint peeling on a solar absorber.

The efficiency of solar collectors is strongly affected by the
quality of the coating used on the absorber.   It is essential that
the absorber coating has good long-term adhesion to other coats and

to the metal and that general paint degradation does not occur [8]. Figure 1.12 shows an example of a paint that did not have these characteristics and eventually peeled from the absorber surface. Since these coats may occasionally experience temperatures above 150°C, very few paints are suitable for this application.   If paint degradation is inevitable, the system must be designed so that the surfaces can be re-painted periodically.

REFERENCES

1. Burrie, K. and Perlin, J., "Solar Water Heaters in Florida 1923-1978", The Co-Evolution Quarterly (Spring 1978) pp.74-79.

2. Ward, J.C. and Lof, G.O.G., "Long-term (18 years) Performance of a Residential Solar Heating System", vol. 18, Solar Energy (1976), pp.301-308.

3. O'Brien, L.F., "Solar Energy Applications in the Design of Buildings", H.J. Cowan, Ed., Applied Sci. Pub., England (1979) pp. 248-261.

4. Boehm, R.F. and Swanson, S.R., "Solar System Design Handbook for Utah", Utah Engineering Experimental Station, Univ. of Utah (June 1978).

5. Balcomb, D., "Passive Solar Handbook", vol. 2, Los Alamos Scientific Labs, DOE/CS-0127/2 (1981).

6. Fuller, W.H., "Solarsoft:- a Design Package for Solar Buildings", Byte (May 1983) pp.426-34.

7. Chopra, P.S., "Reliability and Materials Performance of Solar Heating and Cooling Systems", Argonne Natl. Lab. Report No. SOLAR/0906-79/70 (June 1979).

8. Moore, S.W., "Status Report on Solar Absorber Paint Coatings", Los Alamos Sci. Lab Report LA-8897-SR (July 1981).

CHAPTER 2

Heating-Load Calculations
for Residential Buildings

INTRODUCTION

Unlike conventional heating systems, the main expenses of a solar heating system are the initial purchase and installation costs. Since these costs are closely related to the size (capacity) of the system, economics dictate that these systems be designed to provide only 50 to 75 % of the total heating load (see Chapter 12). This means that during the colder winter months a well designed solar heating system will be fully utilised just providing part of the space heating requirements, and additional backup heating will be needed. For the rest of the year the solar heating system will provide more heat than is needed for space heating. If possible, other uses should be found for the excess energy both to help balance the monthly demand on the system and to help improve the economics of the system. The most common use for part of this excess energy is for domestic hot water heating; however, other uses, such as heat for a swimming pool or a "hot tub" should also be considered.

The space heating load is primarily determined by the rate at which the structure loses heat when the indoor temperature is higher than the outdoor temperature. Heat losses from buildings are mainly due to:

a) Transmission through the confining walls, glass, roof, ceiling, floor and other exposed surfaces.

b) Infiltration of cold outdoor air into the building through cracks and crevices and through open doors and windows and/or outdoor air required for ventilation.

To size the system accurately, a detailed analysis of the space
heating load is needed.   The most commonly used procedures for
calculating heating loads are the ones first recommended by Carrier
[1] and ASHRAE [2].   Computer programs such as CAMEL and TEMPER are
available for the more complex heat load calculations (such as for
multistorey buildings) and allow greater flexibility in the choice
of systems.   For relatively simple structures, such as homes where
it is only necessary to calculate the maximum heating load, a simple
manual calculation will suffice.

In this chapter a month-by-month heating-load analysis will be
performed on a 160 square metre home sited in Sydney using the
methods of references [1] and [3].   The purpose is to demonstrate
the various principles, techniques and assumptions involved in this
type of analysis.   The results will be used in later chapters to
size a solar heating system.

CALCULATION PROCEDURE

If a computer program is not available to perform this analysis, the
calculation procedure described below may give reasonably accurate
predictions.

a)   Obtain the outdoor design temperature either from
      published data (see Table 2.1 [3]) or from the local
      meteorological station.

b)   Select the indoor air temperature to be maintained
      in each room during the coldest weather.   Comfort
      design conditions are usually taken as a dry-bulb
      temperature of $21^{\circ}C$ and a relative humidity of 30%.

c)   Estimate temperatures in adjacent unheated spaces.

d)   Select or calculate the overall heat transmission
      coefficients for all outside surfaces (walls,
      glass, roof, floor, etc.), and inside surfaces
      (internal floors, partitions, ceiling, etc.), if

these are adjacent to unheated spaces (see
Tables 2.2 and 2.3).

e)   Determine the areas of all surfaces through which heat
is expected to be lost.

f)   Calculate the heat loss from all surfaces
exposed to a temperature lower than the heated
space using the following equation:

$$q_c = UA \ (T_i - T_o) \tag{2.1}$$

g)   Calculate the heat required to warm the outside
air that enters the heated space as follows:

$$q_a = 1.196 \ V_a \ (T_i - T_o) \tag{2.2}$$

h)   Sum the heat losses determined in steps f and g.

i)   In cases where a reasonably steady internal heat
release of appreciable magnitude from sources
other than the heating system is available
(lights, people, electric motors, etc.), subtract
this heat release from the sum obtained in item h.
This is especially important for systems of high
initial cost or those for which a demand charge is
based on installed capacity.

EXAMPLE 1: Calculation of the Space Heating Load

Calculate the heat loss at design conditions of the residence shown
in Figure 2.1 which is located in Sydney.   The floor to ceiling
height is 2.5 metres and the occupancy is four persons.   The
building construction is shown in Table 2.4.

Solution:

Considering the entire building,

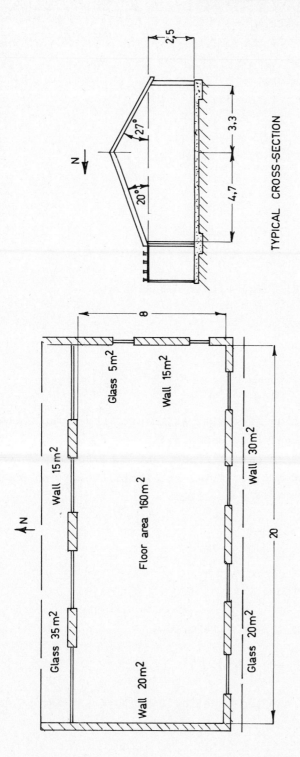

Fig. 2.1.   Building construction dimensions for cooling and heating load estimates.

Design outdoor temperature,

T_o = 7^oC DB   (Table 2.1)

Inside design temperature,

T_i = 21^oC DB   (assumed)

Walls:

q_{cw} = $U_w A_w$ $(T_i - T_o)$, from equation (2.1)

= 2.09 (80) (21-7) = 2340.80 W

Sliding glass doors and windows:

q_{cg} = $U_g A_g (T_i - T_o)$

= 6.42 (60) (21-7) = 5392.80 W

Roof:

q_{cr} = $U_r A_r$ $(T_i - T_o)$

= 0.381 (160) (21-7) = 853.44 W

Floor:

q_{cf} = U_f A_f $(T_i - T_g)$

U_f = $1 / \Sigma R$

for ΣR (Table 2.3)

R_1 = 0.162 (Inside air film, still air, heat flow downwards)

R_2 = 0.500 (Carpet plus underlay)

R_3 = 0.231 (100 mm cellular concrete at 1280 kg/m^3)

R_4 = 0.162 (Outside air film, still air, heat flow downwards)

ΣR = $R_1 + R_2 + R_3 + R_4$ = 1.055

U_f = 1/1.055 = 0.948

T_g = 9^oC   (Table 2.5)

q_{cf} = .948 (160) (21-9) = 1820.16 W

q_c = $q_{cw} + q_{cg} + q_{cr} + q_{cf}$

= 2340.80 + 5392.80 + 853.44 + 1820.16 = 10407.20 W

Cold air entering heated space (from equation (2.2)):

$$q_a = 1.196 \, V_a \, (T_i - T_o)$$

To find V_a for the entire house, use the larger of either the infiltration by the air change method or the air required for ventilation.

Infiltration by the air change method (V_i) (from Table 2.6):

Exposure (exposed)	+ 1/2
Construction (dry)	+ 1/2
Location of windows (2 opposite walls)	+ 1/2
Type of window (weatherstrip)	O
Openable window area per wall area (50% or more)	+ 1/4
Partitioning (heavy)	− 1/2
	1 1/4 ch/h

$$V_i = (ch/h)(\text{volume of space, } m^3)/3.6$$
$$= 1\,1/4 \, (160 \times 2.5)/3.6 = 138.89 \text{ litres/s}$$

Ventilation requirements, V_v
From Table 2.7, use the minimum value of 7 litres/s-person
$$V_v = 4 \times 7 = 28 \text{ litres/s}$$

Mechanical exhausts in residential houses are of an intermittent nature, their effect may be neglected. So since V_i is greater than V_v

$$V_a = V_i = 138.89 \text{ litres/s}$$
and
$$q_a = 1.196 \, (138.89) \, (21-7) = 2325.57 \text{ W}$$

Finally, the heat loss rate ($q_c + q_a$):

$$q_c + q_a = 10407.20 + 2325.57$$
$$= 12732.77 \text{ W or } 12.73 \text{ kW}$$

SPACE HEATING ENERGY CALCULATIONS

The energy required for space heating for a given locality can be calculated by using what is known as the degree-day method. The use of degree-days is based on the concept that on a long term average, solar and internal heat gains will offset losses when the mean daily outdoor temperature is $18.3^{O}C$, and that heating requirements will be proportional to the difference between $18.3^{O}C$ and the mean daily temperature during the heating season.   The monthly average values for 65 locations in Australia are given in Table 2.8 based on $18.3^{O}C$ [4,5].

For well insulated homes where occupants are satisfied with lower inside temperatures, base temperatures as low as $15^{O}C$ have been used (instead of $18.3^{O}C$) as the basis for calculating degree-days.   A lower base temperature will significantly reduce the annual heating requirements for any structure. Table 2.9 [7,8] gives the average monthly degree-day data for the Australian capital cities based on $15^{O}C$.   Table 2.10 [8] gives average annual degree data for various base temperatures for some selected cities in New Zealand.

The general equation for estimating the space heating energy consumption for a specific period (L_s) using degree-days (D) is:

$$L_s = 86400 \ (q_c + q_a)D/(T_i - T_o) \qquad (\text{in J/yr or J/mo}) \quad (2.3)$$

where ($q_c + q_a$) is the space design heating load, T_i and T_o are the inside and outside air temperatures and D is the degree-days per year or per month.

EXAMPLE 2 :   Calculation of Annual Space Heating Load

For the residence described in Example 1, estimate the annual energy required for heating.

Solution:

From equation (2.3),

L_s =   86400 $(q_c + q_a)D/(T_i - T_o)$

$q_c + q_a$ = 12732.77 W (from preceding example problem)

D = 644 degree-days per year (Table 2.8)

T_i = 21°C (assumed)

T_o = 7°C (Table 2.1)

so

L_s =   86400 (12732.77)644/(21 - 7)

   = 5.016 x 10^{10} J/yr (5.06 X 10^4 MJ/yr)

DOMESTIC WATER HEATING LOAD CALCULATIONS

The average domestic water-heating load (L_w) for a household can be calculated from the following equation:-

$$L_w = 1525 \times 10^3 \ (V_w) \ (T_h - T_c), \ \text{J/yr} \qquad (2.4)$$

The mean mains water temperature varies appreciably with location and water source.   For deep wells, the temperature is more or less constant and can be assumed equal to the average yearly ambient temperature.   For surface water and shallow wells, the temperature varies with the outside temperature throughout the year.

The volume of hot water consumed is the most important item in calculating the domestic water heating load.   In existing households this may be readily obtained from charges submitted by the gas or possibly the electric supplier over a period of time. The Sydney-County Council did a survey of 1150 off-peak hot water systems in Sydney and found that the average energy usage for a family ranged from 29 MJ/day in January to 37 MJ/day in July. Table 2.11 contains the results of this survey on hot water consumption and tank sizes for the average Sydney household [6]. However, hot water usage rates vary considerably between households and the type of hot water system also affects the total demand. For example, hot water usage is much greater for systems that operate at mains pressure than for the lower-pressure gravity-feed systems.

EXAMPLE 3 :   Calculation of Monthly Total Heating Load

For the residence in Example 1, estimate the monthly total heating load assuming an average Australian household.

Solution:

$$L_t = L_s + L_w$$
$$L_t = 86400(q_c + q_a)D/(T_i-T_o) + (1525/12)10^3(V_w)(T_h-T_c), \text{ J/mo}$$

where T_c is the average temperature of the water from the mains for that month (Table 2.12 [9]).

$$L_t = 86400 (12732.77)D/(21 - 7) + 127 \times 10^3 (220) (55 - T_c)$$
$$L_t = 7.86 \times 10^7 D + 2.79 \times 10^7 (55 - T_c), \text{ J/mo}$$

or

$$L_t = 78.6 D + 27.9 (55 - T_c) \text{ MJ/mo}$$

A summary of results are given in Table 2.13.

NOMENCLATURE

A = area of wall, glass, roof, ceiling, floor, or any other exposed surface, m^2.

D = degree-days/year or degree-days/mo (Table 2.8, 2.9 and 2.10).

L_s = space heating energy consumption, J/yr or J/mo

L_t = total heating energy load,   J/mo

L_w = energy required for domestic water heating load, J/yr, J/mo

q_a = heat needed to heat the cold outdoor air entering heated space, W.

q_c = design heat loss by conduction and convection through wall, glass, roof, floor and other exposed surfaces, W.

T_c = mean mains water temperature, $^{\circ}C$ (Table 2.12)

T_g = ground temperature, $^{\circ}C$ (Table 2.5)

T_h = hot water supply temperature, usually ranging from 55 to $60^{\circ}C$

T_i = indoor design temperature $^{\circ}C$

T_o = outdoor design air temperature (Table 2.1) or temperature of adjacent unheated space, $^{\circ}C$.

U = overall (air to air) heat transfer coefficient, $W/m^2\ ^{\circ}C$. (Tables 2.2).

V_a = volume of outdoor air entering heated space, litres/s.

V_i = infiltration air, litres/s (Table 2.6).

V_v = ventilation air, litres/s (Table 2.7).

V_w = volume of water used daily, litres/day (Table 2.11).

V_x = exhaust air, litres/s   (Table 2.7).

REFERENCES

1. Carrier International Ltd., Handbook of Air Conditioning Design, McGraw-Hill, New York (1965).

2. ASHRAE Handbook of Fundamentals, Published by the American Society of Heating Refrigeration and Air Conditioning Engineers, Inc., New York (1977).

3. Australian Department of Housing and Construction, Air Conditioning Systems Design Manual, Chapters 1-3, Australian Government Publishing Service, Canberra (1974).

4. Bald, M., Ballinger, J.A. and Hassall, D.N.H., "Performance Evaluation for Passive Solar Heated Buildings", Proceedings of Seminar Developing and Marketing Solar Housing in Temperate Australia, Duntroon, A.C.T. (Nov. 1979).

5. Private communication with M.J. Wooldridge, Division of Mechanical Engineering, CSIRO, Highett, Victoria.

6. Morrison, G.L. and Sapsford, C.M., "Performance of Thermosyphon Solar Water Heaters in Sydney (Final Results)", UNSW School of Mechanical and Industrial Engineering Report 1982/FMT/1.

7. Walsh, P.J. and Spencer, J.W., "Degree-days for Australian Localities", CSIRO, Highett, Vic., 1979, (Unpublished).

8. Keating, E.C., "Average Daily Degree Day Tables, Selected New Zealand Stations", New Zealand Meteorological Service, Government Printer, Wellington, N.Z. (1978).

9. McCaffrey, J.J., "Solar Water Heating for Domestic Use", Solar Energy Systems Course for Master Plumbers, N.S.W. Inst. Tech., NSWIT/PMN/8 (1978).

10. Standards Association of Australia, "Recommended Levels of Thermal Insolation in Dwellings", Draft Standard DR 80111.

TABLE 2.1 Outside Design Conditions – Summer and Winter [3]

LOCATION City or Town & Meteorological Station Locality	SUMMER Comfort or Non-Critical Process Installations °C DB	SUMMER °C WB	WINTER Comfort or Non-Critical Process Installations °C DB	Elevation Above Sea Level m	Latitude Degrees South	Average Daily Range °C DB	Yearly Range °C DB
NEW SOUTH WALES							
Albury	39.5	22.0	1.5	163	36° 06'	16.5	37.5
Casino	39.0	24.0	6.0	24	28° 51'	13.0	33.0
Cessnock	35.0	23.5	7.0	12	32° 54'	13.5	28.0
Cooma	29.5	19.5	-4.0	690	36° 14'	15.0	33.5
Dubbo	39.0	23.0	1.5	265	32° 18'	15.5	37.0
Grafton	35.0	24.0	6.0	6	29° 43'	12.0	29.0
Griffith	35.0	23.0	4.5	128	34° 17'	15.0	30.5
Inverell	38.0	21.5	0.5	604	29° 47'	16.0	37.0
Lismore	35.0	24.0	9.0	11	28° 48'	11.5	26.0
Lithgow	32.0	23.0	1.5	920	33° 30'	14.5	30.5
Maitland	35.0	24.0	6.0	6	32° 44'	13.0	29.0
Moree	40.5	23.5	4.5	209	29° 28'	16.0	36.0
Newcastle	32.0	23.5	4.5	34	32° 35'	6.0	28.0
Orange	35.0	21.5	0.5	868	33° 17'	17.0	34.5
Parkes	38.0	21.5	4.5	316	33° 06'	14.5	33.5
SYDNEY (Mascot airport)	30.5	23.0	7.0	42	33° 51'	7.5	23.5
Tamworth	38.0	21.5	-1.0	378	31° 06'	15.5	39.0
Taree	35.0	24.0	7.0	9	31° 55'	12.0	28.0
Wagga Wagga	39.0	22.0	1.5	187	35° 07'	15.5	37.0
Windsor	36.0	24.0	1.5	12	33° 51'	-	34.5
Wollongong	32.0	23.0	7.0	10	34° 25'	9.0	25.0

TABLE 2.2A Transmission Coefficient U-Veneer Walls [3]

Watts per square metre degree Celsius

CONSTRUCTION		INTERIOR FINISH			
Description		Fibrous Plaster 9 mm (9.9)	Plaster-board 12 mm (11)	Hard-board 6 mm (16)	Vinyl Clad Steel (4.7)
Brick Veneer 90 mm brick with 150 mm air gap	W* S†	2.28 2.20	2.09 2.03	1.90 1.84	- -
110 mm brick with 150 mm air gap	W S	2.19 2.12	2.02 1.96	1.82 1.77	- -
Lightweight Concrete Block Veneer 90 mm block, with 150 mm air gap	W S	1.66 1.62	1.56 1.52	1.45 1.41	- -
140 mm block with 150 mm air gap	W S	1.50 1.47	1.42 1.39	1.33 1.30	- -
Sand & Gravel Aggregate Concrete Block Veneer 90 mm block, with 150 mm air gap	W S	2.11 2.04	1.95 1.90	1.78 1.73	- -
140 mm block, with 150 mm air gap	W S	1.95 1.89	1.81 1.76	1.66 1.62	- -
Weatherboard outside, with timber framing, 100 mm air gap	W S	2.13 2.06	1.97 1.90	1.75 1.72	2.30 2.22
As above with 75 mm of Rockwool & 25 mm air gap	W S	0.523 0.520	0.514 0.510	0.502 0.490	0.536 0.532
Metal clad outside with timber framing 100 mm air gap	W S	2.60 2.48	2.36 2.30	2.09 2.02	2.86 2.74
As above with 75 mm of Rockwool and 25 mm air gap	W S	0.549 0.546	0.540 0.536	0.520 0.518	0.558 0.553

* Assumed still air inside, and wind speed of 7 m/s outside during winter.

† Assumed still air inside, and wind speed of 3.5 m/s outside during summer.

TABLE 2.2B   Transmission Coefficient U-Pitched Roofs

Heat Flow Up

CONSTRUCTION	INTERIOR FINISH							
Description	Plasterboard 12 mm (23)‡		Hardboard 6 mm (27)‡		Caneite			
					18 mm (16)‡		12 mm (14)‡	
	Insulation							
	None	Mineral Fibre 100 mm (6)	None	Mineral Fibre 100 mm (6)	None	Mineral Fibre 100 mm (6)	None	
Concrete Tiles	2.84	0.381	2.48	0.374	1.66	0.348	2.01	
Terracotta Tiles	2.81	0.380	2.46	0.373	1.65	0.347	1.99	
Concrete Tiles and sarking†	1.64	0.347	1.51	0.341	1.17	0.320	1.31	
Terracotta Tiles and sarking†	1.63	0.346	1.50	0.340	1.16	0.319	1.32	

‡ These values include 11 kg/m^2 for the ceiling joists.

† These values are for the case when the reflective sarking is laid on top of the ceiling joists.   If the sarking is placed between the rafters and the tiles with an interior finish of 13 mm gypsum plaster, U = 1.29 W/m^2°C without insulation and U = 0.63, 0.44, 0.36 W/m^2°C without the reflective foil but with mineral fibre insulation thicknesses of 50, 75 and 100 mm, respectively, placed between the ceiling joists [10].

TABLE 2.2C   Transmission Coefficient U-Windows, Skylights, Doors and Glass Block Walls   [3]

Watts per square metre degree Celsius temperature difference

G L A S S

AIR SPACE THICKNESS mm	VERTICAL GLASS						HORIZONTAL GLASS			
	Single		Double		Triple		Single		Double	
	Summer	Winter	Summer	Winter	Summer	Winter	Summer	Winter	Summer	Winter
-	5.89	6.42					4.88	7.95		
5			3.35	3.52	2.31	2.39			2.84	3.98
10			3.15	3.29	2.10	2.16				
15			2.97	3.10	1.93	1.99				
20-100			2.89	3.01	1.88	1.93				

DOORS

THICKNESS OF WOOD mm	U-VALUE	
	Summer	Winter
20	3.27	3.43
30	2.65	2.75
40	2.23	2.30
50	2.01	2.07

HOLLOW GLASS BLOCK WALLS

SIZE	MASS PER UNIT AREA kg/m²	U-VALUE	
		Summer	Winter
196x196x40 mm thick	55	3.00	3.13
196x196x100 mm thick	90	2.89	3.01
196x196x100 mm thick with fibre glass screen dividing the cavity	90	2.69	2.79
300x300x100 mm thick	90	2.79	2.90
300x300x100 mm thick with fibre glass screen dividing the cavity	90	2.57	2.67

TABLE 2.2D   Transmission Coefficient U-Suspended Floors   [1]

Heat Flow Down

Watts per square metre degree Celsius

Still air assumed on both sides

CONSTRUCTION	INTERIOR FINISH		
Description	None	Vinyl Tiles 3 mm (5)	Carpet and Underlay (11)
Concrete: 100 mm + 25 mm sand and cement topping	2.40	2.35	1.22
150 mm + 25 mm topping	2.21	2.17	1.17
300 mm + 25 mm topping	1.80	1.77	1.04
22 mm Timber Floor †	2.10	2.06	1.13
25 mm Chipboard	2.02	1.99	1.11
22 mm Timber floor plus Alfoil with 100 mm air gap	0.845	0.840	0.630
25 mm Chipboard plus Alfoil with 100 mm air gap	0.833	0.828	0.623

† 1120 kg/m^3 has been assumed for timber bearers;

950 kg/m^3 has been assumed for floor boards.

TABLE 2.3 Thermal Resistance R - Building and Insulating Materials [3]

MATERIAL		DENSITY kg/m^3	THICKNESS mm	RESISTIVITY (1/k) m.°C/W	RESISTANCE FOR LISTED THICKNESS m^2. °C/W
AIR SPACES (Temp.diff.15°C)					
Position of Air Space	Direction of Heat Flow				
Horizontal	Up	1.2	100	-	0.150
Horizontal	Down	1.2	100	-	0.217
45° Slope	Up	1.2	100	-	0.155
45° Slope	Down	1.2	100	-	0.180
Vertical	Horizontal	1.2	100	-	0.166
AIR FILMS (Temp.diff. 15°C)					
Still Air:					
Horizontal	Up	1.2	-	-	0.107
Horizontal	Down	1.2	-	-	0.162
45° Slope	Up	1.2	-	-	0.109
45° Slope	Down	1.2	-	-	0.134
Vertical	Horizontal	1.2	-	-	0.120
7 m/s wind: (Winter) any position	any direction	1.2	-	-	0.030
3.5 m/s wind: (Summer) any position	any direction	1.2	-	-	0.044
ALUMINIUM		2675	1.2	4.74x10^{-3}	5.68x10^{-6}
ASBESTOS					
cement board		945	6	5.42	0.0325
cement sheet (fibro cement)		1490	6	3.15	0.0189
BARK FIBRE			-	19.8	-
eucalypt		54	-	22.4	-
redwood		48	-	24.8	-
BITUMEN		1057	-	6.30	-
composition for floors		961	-	1.01	-
emulsion. cement, aggregate		1602	-	1.65	-
pitch mastic, ordinary				1.78	-
BRICK (see also Silica Brick)					
common		1762	90	1.24	0.111
common		1874	90	0.83	0.074
BRICKWORK, common brick wall			90	0.87	0.078
CANE FIBRE (see Sugar Cane Fibre)					
'CANEITE' (pine fibreboard)		256	12	19.3	0.231
CARPET			6	19.8-15.4	0.119-0.09
CARPET UNDERLAY			15	26.7-16.1	0.400-0.242
COCONUT FIBRE, husk		48	-	18.7	-

TABLE 2.3 (Continued) Thermal Resistance R-Building and Insulating Materials

MATERIAL	DENSITY kg/m^3	THICKNESS mm	RESISTIVITY (1/k) m$^2 \cdot °$C/W	RESISTANCE FOR LISTED THICKNESS m$^2 \cdot °$C/W
CONCRETE				
cellular	320	100	12.0	1.196
cellular	480*	100	9.25	0.925
cellular	641*	100	6.93	0.693
clinker aggregate 1:2½:7	1522*	100	3.02	0.302
clinker aggregate 1:2:4	1682*	100	2.48	0.248
clinker aggregate 1:3½:6	1730*	100	1.31	0.131
crushed rock, 1:2:4	2400*	100	0.69	0.069
expanded clay aggregate	801	100	3.47	0.347
gravel, 1:1:2	2339	100	1.07	0.107
slag aggregate, foamed 1:2½:7½	1089*	100	4.08	0.408
COPPER, sheet	8794	1.2	2.6×10^{-3}	3.12×10^{-3}
CORK				
board	144	22	23.91	0.526
granulated, baked	104*	–	5.68	–
raw	160	–	20.4	–
with asphalt or bitumen binder	240	–	18.25	–
with cement binder	280	–	13.87	–
with rubber latex binder	320	–	16.13	–
DIATOMACEOUS EARTH	240-272	–	16.91	–
FELT				
hair	80	–	25.68	–
undercarpet felt	120	–	21.67	–
wool	136-168	–	25.68	–
FIBREBOARD				
(see also WOOD PRODUCTS)	224	12	19.26	0.231
FIBREGLASS (see Mineral Wool)				
GLASS				
cellular slab	144	50	18.25	0.913
cloth, woven	144	–	17.34	–
fibre (see Mineral Wool)				
sheet, window	2515	4	0.95	0.004
sheet, heat resisting	2243	4	0.99	0.004
GYPSUM				
fibrous plaster	1105	9	3.65	0.033
foamed plaster	301	–	16.91	–
plaster	1217	15	2.70	0.041
plasterboard	881	12	5.88	0.071
powder	320	–	15.41	–
JUTE, fibre	36.8	–	34.67	–
LEAD, sheet	11410	1.8	0.029	5.2×10^{-5}
LIMESTONE, Canadian varieties	2547	–	0.78	–
LINOLEUM, inlaid	1297	3	4.62	0.014

TABLE 2.3 (Continued)   Thermal Resistance R-Building and Insulating Materials

MATERIAL	DENSITY kg/m^3	THICKNESS mm	RESISTIVITY (1/k) m^2.°C/W	RESISTANCE FOR LISTED THICKNESS m^2.°C/W
MAGNESIA ASBESTOS, 85%	192	25	16.51	0.413
MAGNESIA CEMENTS	336	-	14.45	-
MANILA HEMP		-	20.40	-
MARBLE, various samples	2643-2804	-	0.77-0.6	-
'MASONITE' (building board from steam-exploded wood)		6	21.0	0.126
MICA, brick		25	6.30	-
MINERAL WOOL				
batts	32	75	31.52	2.364
granulated ('loose fill')	80	-	2.77	-
MORTAR				
cement, sand, 1:3	1890	15	1.14	0.017
cement, sand, 1:4	1954	15	1.08	0.016
PAPER	1089	0.2	7.30	0.002
kraft building paper		0.2	15.41	0.003
PERLITE (see also PLASTER)				
loose expanded granules	32-96	-	23.91	-
cement, sprayed	352	-	12.61	-
PITCH		-	6.94	-
PLASTER (see also GYPSUM)				
foamed	400	-	9.91	-
vermiculite	641	15	4.95	0.074
lime, cement	1442	15	2.10	0.032
lime, sand 1:1		15	2.10	0.032
cement, sand 1:4		15	1.87	0.028
gypsum plaster, sand	1410	15	1.54	0.023
gypsum plaster, perlite	617	15	8.69	0.130
PLYWOOD	529	5	7.22	0.036
fire proofed	561	5	6.60	0.033
POLYSTYRENE				
expanded	16	-	25.68	-
POLYURETHANE				
rigid, foamed, new	24	-	63.04	-
rigid, foamed, aged	24	-	40.79	-
flexible, foamed	40	-	28.90-25.68	-
POLYVINYLCHLORIDE (PVC)				
rigid	1362	-	5.78	-
expanded	9.6	-	34.67	-
PORCELAIN (electrical grade)	2403	-	0.69	-
RUBBER				
cellular slabs	80	50	24.77	1.24
sheet	929	4	6.30	0.025
synthetic	961	4	6.30	0.025
SAND				
building	1506	-	3.30	-
fine silver	1602	-	3.15	-

TABLE 2.3 (Continued)   Thermal Resistance R-Building and Insulating Materials

MATERIAL	DENSITY kg/m^3	THICKNESS mm	RESISTIVITY (1/k) m^2.°C/W	RESISTANCE FOR LISTED THICKNESS m^2.°C/W
SANDSTONE	2002	–	0.77	–
various		–	0.87-0.43	–
SAWDUST	200	–	16.91	–
bonded with urea-formaldehyde resin	440	–	9.91	–
soaked	825	–	2.57	–
bonded with Portland cement 1:2	1201	–	3.47-2.89	–
bonded with Portland cement 1:4	657	–	5.78-4.95	–
Portland cement, sand, sawdust 1:1½:1½	1602	–	1.73-1.39	–
SILICA BRICK	737-1041	–	3.30-1.98	–
SUGAR CANE				
fibre	64	–	24.77	–
fibreboard	216	–	16.13	–
TIMBER				
across grain:				
Balsa	96	25	21.02	0.525
Beech	705	25	5.98	0.150
Deal	609	25	7.97	0.199
Mahogany	705	25	6.42	0.160
Oak	769	25	6.25	0.156
Pitch Pine	657	25	7.22	0.181
Plywood	529	5	7.22	
Plywood, fire proofed	561	5	6.60	
Spruce	416	25	9.50	0.238
Teak	721	25	7.22	0.181
Walnut	657	25	7.22	0.181
Along grain:				
Deal	609	–	4.62	–
Oak	769	–	3.47	–
VERMICULITE				
bonded	208	–	13.34	–
expanded	112-131	–	14.45	–
loose granules	80-112	–	15.41	–
VINYL	1826	3	2.67	0.008
VINYL-ASBESTOS, semi-flexible floor covering	1970	3	1.98	0.0059
WATER	999	–	1.67	–
WOOD PRODUCTS				
fibre and pulp boards (see also Fibreboards)	240	18	18.25	0.328
fibre and pulp boards	320	18	16.91	0.304

TABLE 2.3 (Continued)   Thermal Resistance R-Building and Insulating Materials

MATERIAL	DENSITY kg/m³	THICKNESS mm	RESISTIVITY (1/k) $m^2 \cdot °C/W$	RESISTANCE FOR LISTED THICKNESS $m^2 \cdot °C/W$
fibre and pulp boards	48	18	23.12	0.416
fibre and pulp boards	106	18	22.37	0.403
fibre and pulp boards	144	18	21.67	0.390
fibre and pulp boards	32	18	22.37	0.403
fibre and pulp boards	64	18	23.12	0.416
fibre and pulp boards	96	18	23.91	0.430
fibre and pulp boards	128	18	23.12	0.416
particle board (wood chips bonded with resin)	480	18	9.25	0.166
particle board (wood chips bonded with resin)	641	18	8.16	0.147
particle board (wood chips bonded with resin)	801	18	6.94	0.125
shavings			9.77	
shavings, planer (various woods)	192	–	16.91	–
shredded wood	40	–	17.78	–
shredded wood	101	–	19.26	–
wood wool, acoustical, fluffy	41.6	–	24.77	–
WOOL				
sheep's, low grade	20.8	–	26.67	–
sheep's, low grade	52.9	–	23.12	–

Note: An asterisk (*) appearing in the density
        column signifies that the specimens have
        been conditioned in an atmosphere of
        18°C and 65% relative humidity.

TABLE 2.4   Summary of Construction and Transmission Coefficients

		Area m^2	Description	Mass/Unit Area kg/m^2	U* Summer W/m^2°K	U* Winter W/m °K
G l a s s	N	35	3 mm glass	7.55	5.89	6.42
	S	20	$\tau = 0.86$	7.55	5.89	6.42
	E	5	$\alpha = 0.06$	7.55	5.89	6.42
	W	Nil	$\rho = 0.08$			
W a l l s	N	15	Brick veneer	195	2.03	2.09
	S	30	90 mm brick	195	2.03	2.09
	E	15	150 mm air gap	195	2.03	2.09
	W	20	12 mm plasterboard	195	2.03	2.09
R o o f	N	100	Concrete tiles 12 mm plasterboard 100 mm Mineral	98	0.371	0.381
	S	74	Fibre Insulation	98	0.371	0.381
F l o o r		160	100 mm concrete carpet plus underlay (1280 kg/m^3)	128	1.058	0.948

* assumed still air inside space and 3.5 m/s on the outside

** assumed still air inside space and 7.0 m/s on the outside

Floor: U = 1/ΣR,   where R = Resistance for listed thickness (m^2°K/W)

For 100 mm thick cellular concrete at 1280 kg/m^3 R = 0.231

For wool carpet R = 0.1, for carpet underlay R = 0.4

TABLE 2.5   Ground Temperatures for Estimating Heat Loss through Basement Floors   [1]

Outside Design Temp. °C	-10	-5	0	5	10
Ground Temp. °C	2	4	6	8	10

TABLE 2.6   Infiltration through Windows and Doors - Air Change Method [3]

Parameter	Condition	Ch/h	Condition	Ch/h
Exposure	Sheltered	0	Exposed	+ ½
Construction	Wet	0	Dry	+ ½
Location of Windows	1 wall or 2 adjacent walls	0	2 Opposite walls, 3 or 4 walls	+ ½
Type of window	Weatherstripped	0	Not weatherstripped	+ ½
Openable window area per wall area	Less than 25%	0	50% or more	+ ¼
Partitioning	Nil	0	Heavy	- ½*

*Deductible only if the total air changes considering the other parameters is more than ½.

TABLE 2.7   Ventilation Requirements for Occupants [2]

Application	Required Ventilation Air for Human Occupant	
	Minimum l/s	Recommended l/s
RESIDENTIAL		
Single unit dwellings		
General living areas, bedrooms, utility rooms	2.5	3.5-5
Kitchens, baths, toilet rooms	10	15-25
Multiple unit dwellings and mobile homes		
General living areas, bedrooms, utility rooms	2.5	3.5-5
Kitchens, baths, toilet rooms	10	15-25
Garages	7.5	10-15
COMMERCIAL		
Public restrooms	7.5	10-12.5
General requirements - Merchandising (apply to all forms unless specially noted)		
Sales floors (basement and ground floors)	3.5	5-7.5
Sales floors (upper floors)	3.5	5-7.5
Storage areas (serving sales areas and storerooms)	2.5	3-5.5
Dining rooms	5	7.5-10
Kitchens	15	17.5
Cafeterias, Short-order; Drive-ins, seating areas	15	17.5
Bars (predominantly stand-up)	15	20-25
Cocktail lounges	15	17.5-20
Hotels, motels, resorts		
Bedrooms	3.5	5-7.5
Living rooms (suites)	5	
Baths, toilets (attached to bedrooms)	10	15-25
Corridors	2.5	3.5-5
Lobbies	3.5	5-7.5
Conference rooms (small)	10	12.5-15
Assembly rooms (large)	7.5	10-12.5
Theatres		
Ticket booths	2.5	3.5-5
Lobbies (foyers and lounges)	10	12.5-15
Auditoriums (No smoking)	2.5	2.5-5
Auditoriums (Smoking permitted)	5	5-10
Offices		
General office space	7.5	7.5-12.5
Conference rooms	12.5	15-20
Drafting rooms, art rooms	3.5	5-7.5
Doctors' consulting rooms	5	7.5-10

TABLE 2.7   (Continued)   Ventilation Requirements for Occupants

Application	Required Ventilation Air for Human Occupant	
	Minimum l/s	Recommended l/s
Waiting rooms	5	7.5-10
Computer rooms	2.5	3.5-5
Keypunching rooms	3.5	5-7.5
INSTITUTIONAL		
Schools		
Classrooms	5	5-7.5
Multiple use rooms	5	5-7.5
Laboratories	5	5-7.5
Auditoriums	2.5	2.5-3.8
Gymnasiums	10	12.5-15
Libraries	3.5	5-6
Common rooms, lounges	5	5-7.5
Offices	3.5	5-7.5
Lavatories	7.5	10-12.5
Locker rooms (per locker)	15	20-25
Lunchrooms, dining halls	5	7.5-10
Corridors	7.5	10-12.5
Museums		
Exhibit halls	3.5	5-7.5
Workrooms	5	7.5-10
Warehouses	2.5	3.5-5
Hospitals, Nursing and convalescent homes		
Foyers	10	12.5-15
Hallways	10	12.5-15
Single, dual bedrooms	5	7.5-10
Wards	5	7.5-10
Food service centres	17.5	17.5
Operating rooms, delivery rooms*	10	-
Amphitheatres	5	7.5-10
Physical therapy areas	7.5	10-12.5
Autopsy rooms	15	20-25
Veterinary Hospitals		
Kennels, stalls, operating rooms*	12.5	15-17.5
Reception rooms	5	7.5-10
ORGANIZATIONAL		
Churches, Temples (see theatres, schools & offices)		

*Special requirements or codes may determine requirements

TABLE 2.8   Degree Days of Heating[†] [4]

Station	Jan.	Feb.	Mar.	Apr.	May	June	July	Aug.	Sept.	Oct.	Nov.	Dec.	Year
Adelong	–	–	–	132	263.5	327	365.8	313.1	234	133.3	69	–	1838
Albury	–	–	–	87	232.5	297	347.2	291.4	207	117.8	27	–	1607
Armidale	–	–	12.4	114	257.3	315	368.9	325.5	234	130.2	42	–	1800
Balranald	–	–	–	30	164.3	258	279.0	232.5	159	40.3	–	–	1163
Bathurst	–	–	18.6	138	282.1	327	381.3	334.8	255	158.1	69	–	1964
Bega	–	–	–	84	198.4	267	297.6	254.2	192	105.4	57	–	1456
Bombala	–	5.6	93.0	189	319.3	387	418.5	375.1	294	198.4	141	74.4	2495
Bourke	–	–	–	–	96.1	168	213.9	155.0	48	–	–	–	681
Bowral	–	–	62.0	153	282.1	327	378.2	334.8	261	167.4	126	21.7	2113
Braidwood	–	–	34.1	141	334.8	342	384.4	337.9	252	161.2	105	24.8	2117
Broken Hill	–	–	–	3	148.8	213	251.1	210.8	114	–	–	–	941
Casino	–	–	–	–	31.0	96	139.5	83.7	9	–	–	–	360
Cessnock P.O.	–	–	–	–	130.9	165	226.3	179.8	69	3.1	–	–	764
Charlotte Pass	201.5	182.0	272.8	369	505.3	573	616.9	595.2	516	440.2	369	285.2	4926
Cobar P.O.	–	–	–	–	127.1	201	244.9	186.0	66	–	–	–	825
Coffs Harbour	–	–	–	–	71.3	129	176.7	139.5	69	–	–	–	586
Collarenebri	–	–	–	–	89.9	177	210.8	145.7	48	–	–	–	672
Condobolin P.O.	–	–	–	12	164.3	255	282.1	226.3	135	–	–	–	1075
Cooma North	–	11.2	80.6	195	341.0	396	434.0	393.7	303	210.8	132	68.2	2566
Coonabarabran	–	–	–	72	220.1	285	334.8	288.3	195	74.4	–	–	1470
Cootamundra	–	–	–	78	238.7	297	337.9	291.4	219	108.5	6	–	1577
Cowra	–	–	–	69	217.0	276	319.3	272.8	192	93.0	–	–	1439
Crookwell	–	–	62.0	189	319.3	378	424.7	384.4	303	198.4	135	43.4	2437

TABLE 2.8 (Continued)   Degree Days of Heating

Station	Jan.	Feb.	Mar.	Apr.	May	June	July	Aug.	Sept.	Oct.	Nov.	Dec.	Year
Deniliquin P.O.	-	-	-	45	148.8	261	297.6	244.9	168	58.9	-	-	1224
Dubbo P.O.	-	-	-	9	170.5	234	288.3	241.8	150	27.9	-	-	1122
East Maitland	-	-	-	-	124.0	168	229.4	176.7	72	-	-	-	770
Forbes	-	-	-	51	204.6	270	316.2	266.6	183	65.1	-	-	1357
Glen Innes P.O.	-	-	21.7	117	248.0	312	368.9	325.5	234	136.4	66	3.1	1833
Goulburn	-	-	15.5	132	254.2	318	372.0	328.6	222	155.0	63	-	1860
Grafton Olympic Pool	-	-	-	-	49.6	117	167.4	111.6	48	-	-	-	494
Griffith	-	-	-	60	195.3	267	306.9	257.3	177	65.1	-	-	1329
Hay P.O.	-	-	-	24	167.4	234	337.9	229.4	141	34.1	-	-	1168
Inverell	-	-	-	45	189.1	258	310.0	266.6	177	65.1	-	-	1311
Ivanhoe P.O.	-	-	-	-	158.1	231	257.3	170.5	102	-	-	-	919
Junee	-	-	-	60	213.9	279	316.2	279.0	210	83.7	15	-	1457
Katoomba	3.1	11.2	62.0	153	269.7	324	381.3	344.1	255	176.7	72	40.3	2092
Kiandra Chalet	145.7	137.2	223.2	336	465.0	504	573.5	536.3	447	356.5	270	210.8	4205
Leeton	-	-	-	36	286.0	249	297.6	248.0	168	55.8	-	-	1240
Lismore P.O.	-	-	-	-	52.7	117	164.3	117.8	39	-	-	-	491
Lithgow	9.3	8.4	74.4	174	303.8	357	403.0	365.8	282	189.1	129	52.7	2349
Menindee P.O.	-	-	-	-	124.0	210	226.3	186.0	96	-	-	-	842
Moree P.O.	-	-	-	-	102.3	174	229.4	176.7	69	-	-	-	715
Mudgee	-	-	-	69	204.6	273	325.5	275.9	189	83.7	-	-	1421
Mungindi P.O.	-	-	-	-	77.5	171	210.8	145.7	48	-	-	-	653
Narara (Gosford)	-	-	-	18	136.4	192	235.6	192.2	123	43.4	-	-	941
Narooma	-	-	-	45	148.8	210	238.7	210.8	156	93.0	54	-	1156

TABLE 2.8   (Continued)   Degree Days of Heating

Station	Jan.	Feb.	Mar.	Apr.	May	June	July	Aug.	Sept.	Oct.	Nov.	Dec.	Year
Newcastle	–	–	–	–	83.7	138	170.5	148.8	87	18.6	–	–	647
Nowra	–	–	–	33.0	136.4	192	226.3	204.6	132	65.1	21.0	–	1011
North Wollongong	–	–	–	–	86.8	150	170.5	155.0	75	24.8	6.0	–	668
Orange P.O.	–	–	6.2	129.0	282.1	324	387.5	344.1	252	145.7	54.0	–	1925
Parkes P.O.	–	–	–	18.0	176.7	237	291.4	241.8	159	40.3	–	–	1164
Port Kembla	–	–	–	–	68.2	120	155.0	136.4	81	–	–	–	561
Port Macquarie	–	–	–	–	86.8	141	179.8	155.0	81	31.0	–	–	675
Scone P.O.	–	–	–	15.0	158.1	219	263.5	201.5	111	9.3	–	–	978
Sydney	–	–	–	–	85.0	144	170.0	152.0	81	12.0	–	–	644
Tamworth	–	–	–	3.0	161.2	228	285.2	235.6	144	27.9	–	–	1085
Taree	–	–	–	–	102.3	156	204.6	170.5	96	–	–	–	730
Thredbo Village	136.4	137.2	226.3	312.0	440.2	495	551.8	508.4	414	331.7	282.0	189.1	4024
Tibooburra	–	–	–	–	86.8	168	207.7	158.1	36	–	–	–	657
Wagga	–	–	–	60.0	207.7	282	325.5	272.8	204	83.7	9.0	–	1445
Walgett P.O.	–	–	–	–	102.3	174	223.2	164.3	60	–	–	–	724
Wentworth	–	–	–	27.0	148.8	210	260.4	207.7	117	12.4	–	–	984
Wilcannia P.O.	–	–	–	–	120.9	195	232.5	176.7	42	–	–	–	767
Yass	–	–	15.5	135.0	285.2	345	384.4	334.8	267	158.1	84.0	–	2009
Young	–	–	–	96.0	244.9	309	356.5	313.1	234	117.8	45.0	–	1716

NOTES: †Degree days, or 'accumulated temperatures' below a given base-temperature*, are often used as a guide to average seasonal heating requirements. Heat losses during a particular month can be compared with the degree-days for that period.

*The base temperature used here is 18.3°C. The degree days for a particular month is found by the following method: D.D. = (18.3°C – mean temperature) × No. of days in that month.

TABLE 2.9   Degree Days for Other Localities in Australia and New Zealand*[7,8]

	Jan	Feb	Mar	Apr	May	June	July	Aug	Sep	Oct	Nov	Dec	Year
Sydney Airport M.O.	0	0	0	2	31	73	106	76	34	10	2	0	334
Sydney Regional Office	0	0	0	0	14	47	73	50	18	5	1	0	209
Coffs Harbour (NSW)	0	0	0	0	18	47	82	55	18	3	0	0	223
Canberra A.C.T.	4	3	13	73	201	263	301	253	173	89	38	11	1822
Melbourne R.O.	5	3	7	21	79	128	160	129	78	41	20	11	681
Melbourne (A) Tullamarine	6	2	13	22	95	173	192	167	105	66	31	12	883
Nhill (Vic.)	3	4	10	39	116	117	202	171	116	61	26	12	877
Adelaide (West Terrace)	2	1	2	8	38	76	113	95	53	23	10	5	426
Ceduna S.A.	5	3	7	14	43	79	112	99	52	31	14	8	467
Perth (A) M.O. Guildford	0	0	0	6	20	44	74	70	40	19	4	1	279
Perth Regional Office	0	0	0	2	11	29	55	52	26	12	3	0	189
Brisbane R.O.	0	0	0	0	1	6	22	7	1	0	0	0	37
Alice Springs (A) M.O.	0	0	0	2	31	74	105	61	12	1	0	0	286
Hobart Regional Office	13	10	26	60	142	189	215	191	135	89	55	30	1156
Launceston Tas.	12	10	34	90	186	236	264	231	180	129	74	33	1479
Auckland N.Z.	0	0	1	9	45	97	127	106	66	31	8	2	492
Wellington N.Z.	12	11	24	58	127	183	213	190	141	98	53	25	1135
Christchurch N.Z.	14	14	37	96	190	264	293	252	170	107	56	28	1521
Rotorua N.Z.	6	5	18	65	145	210	235	207	144	90	42	16	1183

*Base temperature = 15°C

TABLE 2.10   Average Annual Totals of Degree-Days for Different Base Temperatures for Localities in New Zealand [8]

Place	Height m	Latitude S	Base Temperature					
			15°C	16°C	17°C	18°C	19°C	20°C
Auckland	49	36° 51'	480	670	890	1140	1420	1730
Rotorua (Whakarewarewa)	307	38° 10'	1200	1460	1750	2060	2400	2750
Gisborne	4	38° 40'	850	1080	1620	1930	2260	2600
Napier	2	39° 30'	860	1080	1330	1600	1900	2220
Palmerston North	34	40° 23'	1070	1330	1610	1920	2250	2600
Wellington (Kelburn)	126	41° 17'	1130	1410	1720	2050	2390	2750
Nelson	2	41° 17'S	1360	1640	1940	2260	2600	2960
Christchurch	30	43° 29'	1530	1820	2120	2440	2770	3110
Ashburton	101	43° 54'	1620	1910	2220	2540	2870	3210
Dunedin Airport	1	45° 56'S	1950	2270	2610	2960	3320	3680
Invercargill	0	46° 25'	2030	2370	2720	3080	3440	3800

TABLE 2.11 Hot Water Demand at 55 C and Tank Sizes [9]

Household No. of Persons	Hot Water Consumption litres/day		Hot Water Tank Size, litres
	Range	Average	
1	60-100	80	140
2	100-160	130	200
3	140-210	180	250
4	180-250	220	310
5	220-280	250	350

TABLE 2.12 Mean Mains Water Temperature (Sydney Data) [6]

Month	Mean Mains Water Temperature °C
Jan.	22
Feb.	21.9
Mar.	20.9
Apr.	18.3
May	15.1
June	12.8
July	11.8
Aug.	13.1
Sept.	15.1
Oct.	17.6
Nov.	19.4
Dec.	21.1
Average	17.43

TABLE 2.13   Monthly Heating Energy Loads Tabulation for Example 3

Month	D	t_c(°C)	L_s (MJ/mo)	L_w (MJ/mo)	L_t (MJ/mo)
Jan.	0	22	0	921	921
Feb.	0	21.9	0	923	923
Mar.	0	20.9	0	951	951
Apr.	0	18.3	0	1024	1024
May	85	15.1	6681	1113	7794
June	144	12.8	11318	1177	12495
July	170	11.8	13362	1205	14567
Aug.	152	13.1	11947	1169	13116
Sep.	81	15.1	6367	1113	7480
Oct.	12	17.6	943	1043	1986
Nov.	0	19.4	0	993	993
DEc.	0	21.1	0	946	946
	Totals (MJ/yr)		50618	12578	63196

CHAPTER 3

Thermal Environment Solar Radiation Data Base, Shading and Energy Conservation

INTRODUCTION

Energy generated in the sun is radiated outwardly in all directions and only about one two-billionth is intercepted by the earth, largely as light and infrared (heat) radiation. The intensity of such radiation normal to the sun's rays just outside the earth's atmosphere (extraterrestrial) at the mean earth-sun distance of 149,638,420 - 80 km is 1353 W/m^2 and is referred to as the "solar constant". Since the distance between the sun and the earth varies during the year, solar radiation ranges from a maximum of 1398.5 W/m^2 on December 21 when the earth is nearest the sun, to a minimum of 1309.9 W/m^2 on June 21 when it is furthest from the sun. This monthly variation in radiation intensity outside the earth's atmosphere (G_o) is given in Table 3.1 [1].

As it enters the earth's atmosphere some of the sun's radiation is reflected and some is absorbed by the air and particles in it such as dust and water vapour. Most of the ultraviolet radiation is absorbed by ozone in the upper atmosphere. Average daily insolation* ranges from 145.2 W/m^2 for most localities in the temperate zone to 532.2 W/m^2 for sunny and arid regions in the tropics.

Since insolation is affected by the local weather patterns, long-term averages of insolation data measured over several years are normally preferred for design calculations. When measured data are not available, the insolation can be estimated from existing

*Intensity of solar radiation on a horizontal surface
at ground level, also known as global radiation.

TABLE 3.1   Extraterrestial Solar Radiation (G_o) and Declination (D) for the
First Day of Each Month.   Base Year 1964   [1]

Month	G_o (W/m^2)	A (J/m^{2}s)	D* (degrees)	Equation of Time (min)	B	C
					Dimensionless Constants	
Jan.	1395.6	1227	+ 20.0	- 11.2	.142	.058
Feb.	1384.3	1211	+ 10.8	- 13.9	.144	.060
Mar.	1363.5	1183	0.0	- 7.5	.156	.071
Apr.	1340.8	1132	- 11.6	+ 1.1	.180	.097
May	1320.6	1101	- 20.0	+ 3.3	.196	.121
June	1309.9	1085	- 23.45	- 1.4	.205	.134
July	1311.1	1082	- 20.6	- 6.2	.207	.136
Aug.	1324.1	1104	- 12.3	- 2.4	.201	.122
Sept.	1344.5	1148	0.0	+ 7.5	.177	.092
Oct.	1366.9	1189	+ 10.5	+ 15.4	.160	.073
Nov.	1387.7	1217	+ 19.8	+ 13.8	.149	.063
Dec.	1398.5	1230	+ 23.45	+ 1.6	.142	.057

*For the northern hemisphere, reverse the sign of the declination D

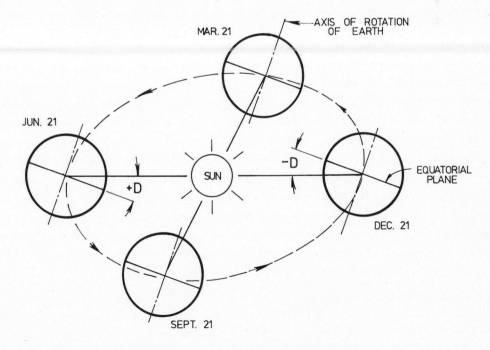

theories.   However, these theories only consider the latitude of
the site and do not accurately account for special local factors.
Thus, the theories would predict the same insolation for foggy San
Francisco as a region in the Nevada desert at the same latitude.

When local insolation data are available, the data are usually
measured on a horizontal plate and must be adjusted for surfaces
that are not horizontal.   These measurements would also not include
such site factors as shading or the radiation reflected from the
ground or from nearby buildings.   A basic understanding of the
theory is needed to allow these factors to be included in any
analysis.   Accordingly, in this chapter the basic theory is covered
before the simplified methods are introduced.   Once the fundamental
theories are developed, the entire analysis can be performed by a
computer.

This chapter contains some simple computer programs (written in
BASIC) which will calculate the various solar angles and the
insolation on an inclined surface from these fundamantal theories.
However, since it is usually necessary to modify these programs to
suit local conditions, or to supply new data, the user should have a
basic understanding of these theories.

THEORETICAL ANALYSIS OF TERRESTRIAL SOLAR IRRADIATION

The amount of solar radiation available during any period is a
significant factor in the design of solar aided systems and in the
determination of heating and cooling loads.   Empirical equations
that involve such quantities as latitude, surface orientation,
declination angle of the sun, solar time, atmospheric absorptivity,
sky diffusion factor, ground reflectance, etc., can be used
(especially with the aid of computer programs) to approximate solar
irradiation at different localities.   These equations are based on
the concept that the total irradiation reaching a terrestrial
surface (G_T) is related to the direct beam radiation normal to the
surface (G_N), the diffuse radiation striking the surface (G_D) and to
the reflected radiation that strikes the surface (G_R).   The
relationship between these quantities is

$$G_T = (G_N) \cos \alpha + G_D + G_R \qquad\qquad (3.1)$$

where the theoretical, clear-sky values for these quantities are given by

$$G_N = A \exp[-B/\sin\theta]$$ (3.1a)

$$G_D = (C)(G_N)(1 + \cos\psi)/2$$ (3.1b)

$$G_R = G_N (C + \sin\theta)(\rho_g)(1 - \cos\psi)/2$$ (3.1c)

and
$$\cos\alpha = \cos\theta \cos\beta \sin\psi + \sin\theta \cos\psi$$ (3.1d)

where ρ is the reflectance of the surroundings (see Table 3.2 [1]), the solar angles α, θ, β and ψ are identified in Figure 3.1 and values of A, B and C in Table 3.1.

TABLE 3.2   Solar Reflectances of Various Foreground Surfaces [1]

Type of Surface	Incidence Angle, degree					
	20	30	40	50	60	70
New concrete	.31	.31	.32	.32	.33	.34
Old concrete	.22	.22	.22	.23	.23	.25
Bright green grass	.21	.22	.23	.25	.28	.31
Crushed rock	.20	.20	.20	.20	.20	.20
Bitumen and gravel roof	.14	.14	.14	.14	.14	.14
Bituminous parking lot	.09	.09	.10	.10	.11	.12

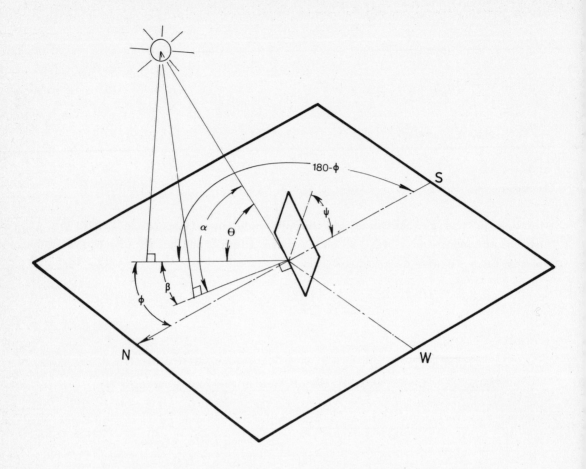

α   =   Incidence angle referred to a line normal to the surface,
        degrees (Figure 3.1)

β   =   Surface azimuth angle referred to a line normal to the
        intersection between the surface and the horizontal
        plane,

θ   =   Solar altitude angle
        degrees (Figure 3.1)

φ   =   Solar azimuth angle, degrees (Figure 3.1)

ψ   =   Surface tilt angle, degrees (Figure 3.1)

Fig. 3.1. Solar angles for a northerly oriented vertical surface.

Irradiation values calculated from equation (3.1) are for a typical cloudless sky and should be adjusted accordingly if clear dry skies or cloudiness predominate.   Consequently, computed irradiation values can depart significantly from actual values unless local conditions, like cloud cover and humidity, are considered.

A more accurate method of obtaining on-site irradiation is by direct measurement.   This is done by the use of an instrument called a pyranometer which produces a voltage related to the difference in temperature between a black and a white (or shaded) surface when exposed to solar radiation.   This method is more tedious than the theoretical procedure used here since accurate data can only be obtained if taken as an average of measurements made over long periods of time (5 years or more) making it practical only for metropolitan areas and other centres of population where there are other needs for accurate irradiation data.   Often, measured solar insolation data may be obtained from local meteorology stations. With more and more emphasis being placed on solar energy as an alternative energy source, education and research institutions have taken the cue with research directed towards obtaining more accurate and extensive records of terrestrial irradiation.

Table 3.3 [2] gives the measured monthly average insolation for 12 localities in Australia.   The equivalent irradiation on an inclined surface can be calculated from these insolation data using trigonometry and assumptions involving the ratio of direct to diffuse radiation.   Table 3.4 [3] gives the computed irradiation values for north oriented surfaces located in Sydney which are inclined at angles from 0 to 90 degrees.   These calculations used hourly global radiation data for horizontal surfaces that had been measured between 1972 and 1976.   Another calculation procedure, developed by Liu and Jordan [4], will be described in the next section.

Some other sources of solar data for Australia are given in Appendix 1.

TABLE 3.3   Measured Global Radiation Input (MJ m^{-2}day^{-1}) for 12 Australian Stations for Each Month [2]

Horizontal	Jan.	Feb.	Mar.	Apr.	May	June	July	Aug.	Sept.	Oct.	Nov.	Dec.	Ave. Year
Adelaide	24.97	22.61	18.05	13.33	9.95	7.93	8.94	11.81	16.87	19.57	23.28	25.14	16.87
Alice Springs	26.73	25.44	23.85	19.31	15.67	14.76	16.01	19.87	22.94	25.10	26.23	26.35	21.79
Brisbane	25.33	23.06	20.22	15.88	12.47	11.53	12.47	15.12	18.71	22.11	24.76	25.14	18.90
Canberra	25.89	23.28	18.51	14.42	11.02	8.63	9.54	12.72	17.83	21.12	24.87	26.69	17.85
Darwin	18.40	18.62	19.65	18.40	19.65	19.19	19.31	21.80	23.05	23.17	21.35	19.87	20.24
Hobart	23.19	19.94	15.13	10.61	6.65	5.09	5.51	8.48	13.15	17.96	20.79	23.19	14.14
Longreach	28.73	24.07	22.83	19.99	15.78	15.10	16.35	18.51	22.83	26.12	28.84	27.03	22.17
Melbourne	24.87	21.46	16.69	11.58	7.57	6.25	6.93	9.65	13.17	18.17	21.92	24.19	15.15
Perth	27.25	25.21	20.67	14.88	10.56	9.43	10.33	13.63	17.60	22.48	26.57	28.84	18.91
Port Hedland	27.59	24.76	25.21	19.99	15.10	16.47	17.72	21.01	23.62	26.80	28.73	29.30	23.02
Sydney	22.48	18.96	18.28	18.51	10.67	8.97	10.45	13.06	16.69	21.46	24.53	23.17	17.26
Townsville	21.69	20.55	20.10	18.62	16.13	15.78	16.35	19.42	23.73	23.73	25.66	25.44	20.60

TABLE 3.4   Mean Daily Irradiation for an Inclined Surface Located at Kensington, Sydney [3]

Inclination Degrees	Jan.	Feb.	Mar.	Apr.	May	June	July	Aug.	Sept.	Oct.	Nov.	Dec.	Ave.
0.0	21.9	18.9	16.1	13.4	10.5	8.1	10.3	12.1	16.5	18.6	22.5	24.2	16.0
17.0	21.2	19.0	17.0	15.4	13.2	10.4	13.3	14.2	18.1	19.1	22.2	23.5	17.2
34.0	19.6	18.1	17.0	16.5	15.0	12.1	15.4	15.5	18.7	18.5	20.6	21.5	17.3
50.0	17.0	16.3	16.2	16.6	15.7	12.9	16.4	15.8	18.1	17.0	18.2	18.6	16.5
70.0	13.0	13.0	13.9	15.3	15.3	12.8	16.2	14.9	16.0	14.0	14.1	14.0	14.3
90.0	8.5	9.0	10.5	12.6	13.3	11.4	14.3	12.6	12.6	10.1	9.5	9.2	11.1

SOLAR ANGLES

Knowledge of the path of the sun as it moves across the sky is essential in determining the solar heat gain for the structure, the best house orientation and the best locations for windows, skylights and solar collectors.

The position of the sun at any time along its path can be related to a ground observer (Figure 3.1) by the altitude and azimuth solar angles which can be calculated as follows:

$$\sin \theta \;\; = \;\; \cos L \cos D \cos H + \sin L \sin D \qquad\qquad (3.2)$$

$$\sin \phi \;\; = \;\;\; \cos D \sin H / \cos \theta \qquad\qquad\qquad (3.3)$$

where

$$H = (\text{Solar time} - 12) \, 60/4 \text{ (Use absolute value) } (3.3a)$$

Solar time (or sun time) depends on the position of the sun and could be different for localities in the same clock time zone. Solar noon occurs when the sun is exactly midway between sunrise and sunset.   The following examples show the steps necessary to convert local time to sun time and to calculate the solar angles for an exposed surface.

EXAMPLE 1 :   Conversion of Clock Time to Solar (Sun) Time

Find the solar time for a building located at 35°S latitude and 151°E longitude (Sydney) on December 22 when the clock time is   11 a.m.

Solution:

Since daylight saving time is in effect in Sydney on December 22, subtract one hour from clock time.   Next, a correction of four minutes for every degree of longitude by which the location departs from the standard time meridian must be applied.   (For Australia the standard meridians are: 150° for Queensland, New South Wales and Victoria; 142.5^O for the Northern Territory and South Australia; and 120° for Western Australia).   If the location is east of the standard meridian, this correction is added to the clock time and if the location is west of the standard this correction is subtracted.

Finally, add the "Equation of Time" correction (see Table 3.1). This is not actually an equation but rather a correction factor that allows for the earth's elliptical orbit and tilted axis.   Using this procedure, the local solar time is,

    solar time = local standard time
        + (local latitude - local standard meridian) 4/60
        + "Equation of Time"
    or, for 11 a.m. in Sydney on December 22

    solar time = (11 - 1) + (151 - 150) 4/60 + 1.6/60

                = 10.093 hrs (10:056 a.m.)

EXAMPLE 2 :   Solar Angles

A 30 m tall building is sited at the same location used in Example 1. Find the solar altitude (θ) and azimuth angle (ϕ) at the given site using equations 3.2, 3.3, and 3.3a and compare these results with the values obtained from Table 3.5 [5].

Solution:
    $\sin \theta = \cos L \cos D \cos H + \sin L \sin D$

    H = (solar time - 12) 60/4
    H = (10.093 - 12) 60/4 = 28.6^O (absolute value)

   $\sin \theta = \cos 35^O \cos 23.45^O \cos 28.6^O + \sin 35^O \sin 23.45^O$
        = 0.888
or
        θ = $\underline{62.63^O}$ ans.

TABLE 3.5   Solar Altitude and Azimuth Angles   [5,6]

LATITUDE 10° SOUTH

Sun Time	Jan.21 Alt.	Az.	Feb.20 Alt.	Az.	Mar.22 Alt.	Az.	Apr.20 Alt.	Az.	May 21 Alt.	Az.	Jun.21 Alt.	Az.	Jul.23 Alt.	Az.	Aug.24 Alt.	Az.	Sep.22 Alt.	Az.	Oct.23 Alt.	Az.	Nov.21 Alt.	Az.	Dec.22 Alt.	Az.
6 am	3	110	2	101															2	101	3	109	4	113
7	17	108	16	99	15	87	12	76	10	67	9	64	10	67	12	76	15	87	16	99	17	108	18	112
8	31	108	31	97	29	84	27	72	24	63	22	59	24	63	27	72	29	84	31	97	31	108	31	111
9	46	109	46	96	44	80	40	66	37	56	35	52	36	55	40	66	44	79	46	96	46	108	45	113
10	59	113	60	95	58	73	53	55	48	44	46	41	48	44	53	55	58	72	60	95	59	112	58	119
11	72	127	75	96	72	56	64	35	57	26	53	24	56	26	64	35	72	55	75	96	73	126	70	135
12N	80	180	89	180	80	0	69	0	60	0	57	0	60	0	69	0	79	0	89	180	80	180	77	180
1 pm	72	233	75	264	72	304	64	325	57	334	53	336	56	334	64	325	72	305	75	264	73	234	70	225
2	59	247	60	265	58	287	53	305	48	316	46	319	48	316	53	305	58	288	60	265	59	248	58	241
3	46	251	46	264	44	280	40	294	37	304	35	308	36	305	40	294	44	281	46	264	46	252	45	247
4	31	252	31	263	29	276	27	288	24	297	22	301	24	297	27	288	29	276	31	263	31	252	31	249
5	17	252	16	261	15	273	12	284	10	293	9	296	10	293	12	284	15	273	16	261	17	252	18	248
6	3	250	2	259															2	259	3	251	4	247

LATITUDE 20° SOUTH

Sun Time	Jan.21 Alt.	Az.	Feb.20 Alt.	Az.	Mar.22 Alt.	Az.	Apr.20 Alt.	Az.	May 21 Alt.	Az.	Jun.21 Alt.	Az.	Jul.23 Alt.	Az.	Aug.24 Alt.	Az.	Sep.22 Alt.	Az.	Oct.23 Alt.	Az.	Nov.21 Alt.	Az.	Dec.22 Alt.	Az.
6 am	7	109	4	101															4	100	7	109	8	112
7	20	105	18	96	14	84	10	74	6	66	5	63	6	66	10	74	14	84	18	96	20	104	21	108
8	34	101	32	91	28	78	23	68	19	59	17	56	19	59	23	67	28	78	32	91	34	101	35	105
9	48	98	46	85	41	71	36	59	30	50	28	47	30	50	36	59	41	70	46	85	48	98	48	103
10	62	95	60	78	54	59	47	46	40	38	38	35	40	38	47	46	54	58	60	77	62	95	62	103
11	76	93	73	61	65	38	55	27	47	21	44	19	47	21	55	26	65	37	73	61	76	92	76	107
12N	90	180	81	0	70	0	59	0	50	0	47	0	50	0	59	0	69	0	81	0	90	0	87	180
1 pm	76	267	73	299	65	322	55	333	47	339	44	341	47	339	55	334	65	323	73	299	76	268	76	253
2	62	265	60	282	54	301	47	314	40	322	38	325	40	322	47	314	54	302	60	283	62	265	62	257
3	48	262	46	275	41	289	36	301	30	310	28	313	30	310	36	301	41	290	46	275	48	262	48	257
4	34	259	32	269	28	282	23	292	19	301	17	304	19	301	23	293	28	282	32	269	34	259	35	255
5	20	255	18	264	14	276	10	286	6	294	5	297	6	294	10	286	14	276	18	264	20	256	21	252
6	7	251	4	259															4	260	7	251	8	248

TABLE 3.5 (Continued)   Solar Altitude and Azimuth Angles

LATITUDE 30° SOUTH

Sun Time	Jan.21		Feb.20		Mar.22		Apr.20		May 21		Jun.21		Jul.23		Aug.24		Sep.22		Oct.23		Nov.21		Dec.22	
	Alt.	Az.	Alt.	Az.	Alt.	Az.	Alt.	Az.	Alt.	Az.	Alt.	Az.	Alt.	Az.	Alt.	Az.	Alt.	Az.	Alt.	Az.	Alt.	Az.	Alt.	Az.
6 am	10	108	6	100															6	100	10	107	11	111
7	22	101	18	92	13	82	7	73	2	65	0	62	2	65	7	73	13	82	18	92	22	101	24	104
8	35	94	31	85	25	74	19	64	14	57	11	54	13	57	19	64	25	73	31	85	35	94	37	98
9	48	87	44	76	38	63	30	53	24	47	21	44	24	46	30	53	37	63	44	75	48	87	50	92
10	61	77	56	62	48	49	40	40	32	34	29	32	32	34	40	39	48	48	56	62	61	76	62	83
11	73	57	67	40	56	28	46	22	38	18	35	17	38	18	46	21	56	28	67	40	73	56	75	67
12N	80	0	71	0	60	0	49	0	40	0	37	0	40	0	49	0	59	0	71	0	80	0	83	0
1 pm	73	303	67	320	56	332	46	338	38	342	35	343	38	342	46	339	56	332	67	320	73	304	75	293
2	61	283	56	298	48	311	40	320	32	326	29	328	32	326	40	321	48	312	56	298	61	284	62	277
3	48	273	44	284	38	297	30	307	24	313	21	316	24	314	30	307	37	297	44	285	48	273	50	268
4	35	266	31	275	25	286	19	296	14	303	11	306	13	303	19	296	25	287	31	275	35	266	37	262
5	22	259	18	268	13	278	7	287	2	295	0	298	2	295	7	287	13	278	18	268	22	259	24	255
6	10	252	6	260															6	260	10	253	11	249

LATITUDE 40° SOUTH

Sun Time	Jan.21		Feb.20		Mar.22		Apr.20		May 21		Jun.21		Jul.23		Aug.24		Sep.22		Oct.23		Nov.21		Dec.22	
	Alt.	Az.	Alt.	Az.	Alt.	Az.	Alt.	Az.	Alt.	Az.	Alt.	Az.	Alt.	Az.	Alt.	Az.	Alt.	Az.	Alt.	Az.	Alt.	Az.	Alt.	Az.
6 am	13	106	7	99															7	99	13	105	15	108
7	24	97	19	89	11	80	4	72							4	72	11	80	19	89	24	96	26	100
8	35	87	30	79	22	69	14	61	8	55	5	53	8	55	14	61	22	69	30	79	35	87	37	90
9	47	76	41	67	33	57	24	49	17	44	14	42	17	44	24	49	32	57	41	67	47	76	49	80
10	58	61	51	51	41	42	32	35	24	31	21	29	24	31	32	35	41	41	51	51	57	61	60	66
11	66	37	58	29	47	22	37	19	28	16	25	15	28	16	37	18	47	22	58	29	66	37	69	42
12N	70	0	61	0	50	0	39	0	30	0	27	0	30	0	39	0	49	0	61	0	70	0	73	0
1 pm	66	323	58	331	47	338	37	341	28	344	25	345	28	344	37	342	47	338	58	331	66	323	69	318
2	58	299	51	309	41	318	32	325	24	329	21	331	24	329	32	325	41	319	51	309	57	299	60	294
3	47	284	41	293	33	303	24	311	17	316	14	318	17	316	24	311	32	303	41	293	47	284	49	280
4	35	273	30	281	22	291	14	299	8	305	5	307	8	305	14	299	22	291	30	281	35	273	37	270
5	24	263	19	271	11	280	4	288							4	288	11	280	19	271	24	264	26	260
6	13	254	7	261															7	261	13	255	15	252

$$\sin \phi = [\cos D \sin H]/\cos \theta$$
$$= [\cos 23.45 \sin 28.6]/\cos 62.63$$
$$= 0.9552$$

and

$$\phi = \underline{72.79^{\circ}} \text{ ans}$$

Interpolating from Table 3.5

$$\theta = (63.21 + 60.8)/2 = 62.02^{\circ} \text{ as compared with } 62.63^{\circ}$$

$$\phi = (81.51 + 63.77)/2 = 72.64^{\circ} \text{ as compared with } 72.79^{\circ}.$$

A less tedious method of determining these solar angles as well as the solar time is to use a computer program such as the one listed below. This program is written in BASIC and is a simplified version of one of the Apple$^{\copyright}$ programs on the SOLARAUST disk available from Pregamon Press.

COMPUTER PROGRAM FOR SOLAR ANGLES

```
500  DIM DE(12): DIM AI(12): DIM BB(12): DIM CC(12)
600  DIM ET(12)
1000 PRINT "WHAT IS THE ANGLE OF INCLINATION OF THE SURFACE WITH
RESPECT TO THE HORIZONTAL PLANE (IN DEGREES)?": INPUT TILT
1010 PI = 3.14159/180
1015 PFI = TILT*PI
1020 PRINT "WHAT IS YOUR LATITUDE?":INPUT LA
1025 PRINT "WHAT IS YOUR LONGITUDE?":INPUT LO
1030 PRINT "FOR WEST LONGITUDES TYPE [-1], FOR EAST LONGITUDES TYPE
[1]":INPUT ZZ
1040 PRINT "WHAT IS YOUR LOCAL TIME (ON A 24 HOUR CLOCK)?": INPUT
LT
1045 PRINT "DOES YOUR LOCAL TIME INCLUDE DAYLIGHT SAVINGS? YES = 1,
NO = 0": INPUT DS
1050 PRINT " WHAT IS YOUR LOCAL STANDARD MERIDIAN?": INPUT SM
1055 DATA -20, -10.8,  0,11.6,  20,  23.45,  20.6,  12.3,  0,  -10.5,
-19.8, -23.45
```

```
1060 FOR I = 1 TO 12 STEP 1: READ DE(I)
1070 NEXT
1080 DATA 1227, 1211, 1183, 1132, 1101, 1085, 1082, 1104, 1148,
1189, 1217, 1230
1090 FOR I = 1 TO 12 STEP 1: READ AI(I)
1095 NEXT
1100 DATA .142, .144, .156, .180, .196, .205, .207, .201, .177,
.160, .149, .142
1110 FOR I = 1 TO 12 STEP 1: READ BB(I)
1120 NEXT
1130 DATA .058, .060, .071, .097, .121, .134, .136, .122, .092,
.073, .063, .057
1140 FOR I = 1 TO 12 STEP 1: READ CC(I)
1150 NEXT
1155 DATA -11.2, -13.9, -7.5, 1.1, 3.3, -1.4, -6.2, -2.4, 7.5, 15.4,
13.8, 1.6
1160 FOR I = 1 TO 12 STEP 1: READ ET(I)
1170 NEXT
1800 PRINT "ENTER THE NUMBER OF THE MONTH BEING ANALYSED"
1810 PRINT "(JANUARY = 1 AND DECEMBER = 12)": INPUT J
1820 ST = LT - DS + ZZ* (LO - SM)/15 + ET(J)/60
1830 AH = ABS((ST - 12)*15*PI)
1840 PRINT "TILT ANGLE = "; TILT;",    LATITUDE = ";LA;",    SOLAR
TIME = ";ST
1850 X = COS(LA * PI) * COS(DE(J) * PI) * COS(AH) + YY * SIN(LA *
PI) * SIN(DE(J) * PI)
1860 THETA = ATN(X/SQR(-X*X + 1))
1870 X = COS(DE(J) * PI)) * SIN(AH)/COS(THETA)
1880 PSI = ATN(X/SQR(-X * X + 1))
1890 X = COS(THETA) * COS(BETA) * SIN(PSI) + SIN(THETA) * COS(PSI)
1900 ALPHA = - ATN(X/SQR(- X * X + 1)) + 1.5708
1910 TA = THETA/PI
1920 PA = PSI/PI
1930 AA = ALPHA/PI
1940 PRINT "    THETA = ";TA;"    PSI = ";PA; "    ALPHA = ";AA
1950 REM THETA, ALPHA AND PSI ARE THE SOLAR ANGLES FOR AN INCLINDED,
NORTH-FACING PLATE IN THE SOUTHERN HEMISPHERE
1960 END
```

SHADING

The direct gain of solar radiation through windows and doors can represent a significant part of the heating or cooling load, so shading devices are often used. Computer programs are available for sizing shading devices and calculating the shade on a site throughout the year caused by nearby structures. Unfortunately, these programs require large computers to run. Graphical techniques and specially designed hand-held instruments are often used in practice to help select shading devices or to predict the shading patterns on a particular site. As the use of microcomputers become more widespread new computer programs that perform shading calculations will certainly be developed for use with these systems. The calculation procedure used in this section will demonstrate the basic principles on which such a computer program could be based.

In addition to the shape and position of a structure, it is also necessary to determine the solar angles at different times of the year. As was seen in the previous section, these solar angles can be either calculated from the fundamental equations, be obtained from the data in Table 3.5, or be calculated with the computer program listed above. Solar angles can be used to determine the solar radiation intensity on horizontal and inclined surfaces, or the shading produced by reveals, by overhangs, by fins and by adjacent buildings. Example 3 shows how the shading from a window reveal (S_r) and the shading from a window overhang (S_o) can be obtained from the geometry of the structure.

EXAMPLE 3 : Shading Calculations

Find the shadow on a window in Sydney at 10.056 a.m. on December 21. The window is on a north wall with the reveals and overhang as shown in the figures.

Solution

Shading from reveals,

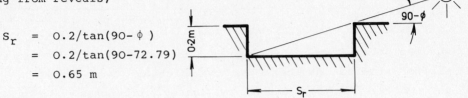

$$S_r = 0.2/\tan(90-\phi)$$
$$= 0.2/\tan(90-72.79)$$
$$= 0.65 \text{ m}$$

Shading from overhang,

$$S_O = (0.2 + 0.6) \tan\theta' - 0.3$$

from the sun-angle figure,

$$\tan\theta' = \tan\theta/\cos\beta$$
so
$$S_O = (0.8)[\tan\theta/\cos\beta] - 0.3$$

For a North-facing surface, $\beta = \phi$, so

$$S_O = (0.8)[\tan 62.63/\cos 72.79] - 0.3$$
$$= 4.94 \text{ m}$$

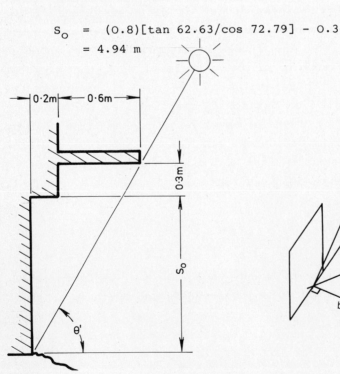

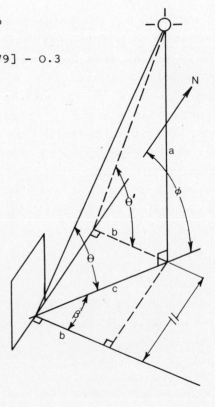

The chart in Figure 3.2 can provide an easy way of estimating the amount of shading from vertical and horizontal structures.   As an example, consider the last problem.

Shading from the reveals:
   Azimuth angle = 72.79°
   Orientation   = North
   From Figure 3.2, shading from side = 3.2 m/m
so, S_r = (0.2)(3.2) = 0.64 m (which is close to the
   value calculated above)

Shading from overhang:
   Altitude angle = 62.63°
   From Figure 3.2, the shading from the top = 6.5 m/m
so, S_o = (0.2 + 0.6)(6.5) − 0.3 = 4.9 m

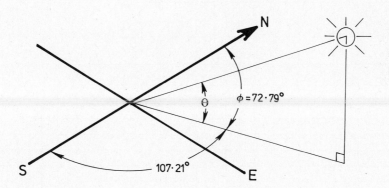

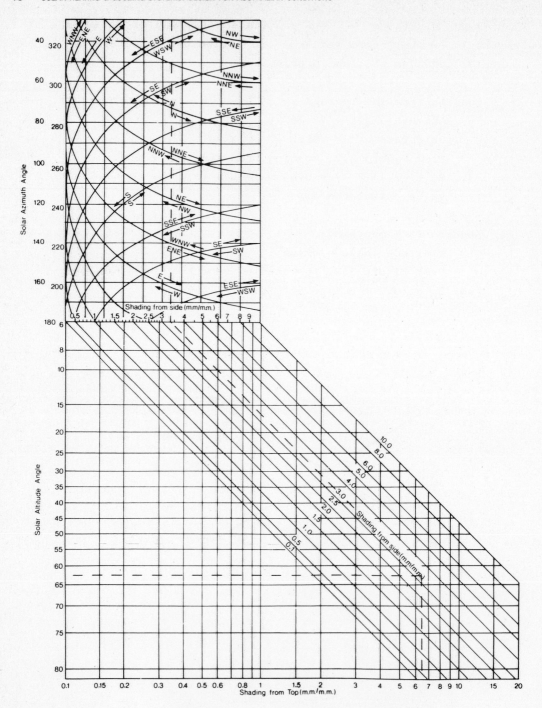

Fig. 3.2.   Shading from reveals and overhangs.
(Permission from McGraw-Hill Book Company [6].)

MONTHLY INSOLATION ON A TILTED SURFACE

The calculation procedure developed by Liu and Jordan [4] allows
insolation measured on a horizontal surface (e.g. Table 3.3) to be
used to calculate the insolation on tilted surfaces, such as on a
roof or on a solar collector. Liu and Jordan introduced a
dimensionless "cloudiness index" (K_T) which they defined as

$$K_T = G/G_O \qquad\qquad (3.4)$$

where G is the daily average insolation per month measured on a
horizontal surface and G_O is the daily average extraterrestrial
insolation per month from Table 3.1. A large value of K_T indicates
clear conditions and a small value indicates cloudy conditions.
Normally, K_T will lie between 0.3 and 0.8.

Since extraterrestrial solar radiation (G_O) given in Table 3.1 is
for solar noon on a particular day of each month, the monthly
average (G_{Om}) must be calculated from geometric considerations.
The relationship for G_{Om} is

$$G_{Om} = (24\ G_O/\pi)\ [1 + 0.033\cos(0.97297\ N)]\ [\cos(L)\cos(D)$$
$$\cos(w_s) + w_s \sin(L)\sin(D)] \qquad (3.5)$$

where L is the latitude, D is the declination, w_s is the sunset hour
angle (in radians) for a horizontal surface and N is the number of
the day of the year, usually taken as the 16th day of each month.
For level sites that are not shaded by hills or trees, the sunset
hour angle is given by

$$\cos(w_s) = -\tan(L)\tan(D) \qquad\qquad (3.6)$$

For north-facing tilted surfaces, the calculation procedure uses a
tilt factor (R_D) which is the ratio of the average daily beam
radiation on the tilted surface to that on a horizontal surface for
each month. This ratio is given by

$$R_D = [\cos(L- \psi)\cos D \cos w_s + w_s \sin(L- \psi)$$
$$\sin D]/[\cos L \cos D \sin w_s \sin L \sin D] \qquad (3.7)$$

The total insolation on a north-facing tilted surface (G_T) is the sum of this direct beam radiation, the diffuse radiation and the reflected radiation that strike the tilted surface. The relationship for this total insolation is [4]

$$G_T = R \times G \qquad (3.8)$$

where

$$R = (1 - D_H)R_D + D_H(1 + \cos \psi)/2 + \quad g(1 - \cos \psi)/2 \quad (3.9)$$

and

$$D_H = 1 - 1.015 \, K_T \qquad (3.10)$$

This relationship for D_H is not the one originally recommended by Liu and Jordan, but one that has been found to predict winter values of R to within 2% for Australia.

The simple computer program (written in BASIC) listed below can be used on small personal computers to calculate the total insolation on north-facing tilted surfaces using the method described above. This program is a simplified version of one of the Apple[©] programs used on the SOLARAUST disk available from Pergamon Press.

INSOLATION ON TILTED SURFACES

```
100   REM THIS PROGRAM WILL CONVERT THE AVERAGE MONTHLY
      INSOLATION DATA FOR A HORIZONTAL SURFACE TO THE VALUE
      APPLICABLE FOR AN INCLINED SURFACE

110   DIM AI(12): DIM DN(12): DIM RAD(12): DIM DAL(12): DIM
      CA(12): DIM CW(12): DIM XI(12): DIM XJ(12): DIM ISC(12)
```

120   PRINT "TYPE IN AVERAGE DAILY INSOLATION DATA MEASURED ON A
       HORIZONTAL SURFACE AT YOUR SITE FOR EACH MONTH STARTING WITH
       JANUARY"

130   FOR I = 1 TO 12 STEP 1: INPUT RAD(I): NEXT
140   DATA 1395.6, 1384.3, 1363.5, 1340.8, 1320.6, 1309.9, 1311.1,
1324.1, 1344.5, 1366.9, 1387.7, 1398.5
150   FOR I = 1 TO 12 STEP 1: READ ISC(I): NEXT
160   PRINT"WHAT IS THE ANGLE OF INCLINATION OF YOUR SURFACE WITH
RESPECT TO THE HORIZONTAL PLANE [IN DEGREES]?": INPUT TILT
170   PI = 3.14159/180
180   PRINT "WHAT IS YOUR LATITUDE?":INPUT LA
190   PRINT "FOR THE NORTHERN HEMISPHERE TYPE [1], FOR THE SOUTHERN
HEMISPHERE TYPE [-1].": INPUT YY
200   DATA 16, 50, 76, 106, 136, 169, 198, 228, 261, 288, 320, 354
210   FOR I = 1 TO 12 STEP 1: READ DN(I): NEXT
220   PRINT"MONTHLY RADIATION CALCULATED FOR A SURFACE TILTED AT
       ";TILT;"   DEGREES   AT   THE   SAME   LOCATION   [STARTING   WITH
JANUARY]"
230   FOR J = 1 TO 12 STEP 1
240   DA = YY * 23.45 * SIN((DN(J) - 80) *.97297 * PI)
250   CA = -TAN(LA * PI) * TAN(DA * PI)
260   XJ = (- ATN(CA/SQR(- CA * CA + 1)) + 1.5708)/PI
270   Y = COS(LA * PI) * COS(DA * PI) * SIN(XJ * PI) + PI * XJ *
       SIN(LA * PI) * SIN(DA * PI)
280   RD = (COS((LA - TILT) * PI) * COS(DA * PI) * SIN(XJ * PI) + PI
       * XJ * SIN((LA - TILT) * PI) * SIN(DA * PI))/Y
290   IO = ISC(J) * (1 + 0.033 * COS(DN(J) * 0.97297 * PI)) * .0036
300   KT = RAD(J) * 3.14159/(24 * IO * Y)
310   DH = 1 - 1.015 * KT
320   RB = (1 - DH) * RD + DH * (1 + COS(TILT*PI))/2 + .075 * (1 -
       COS(TILT * PI))
330   HT = RB * RAD(J)
340   PRINT HT;",   ";
350   NEXT
360   END

NOMENCLATURE

D = Declination or angular distance of the sun (north or south) from the celestial equator, degrees (Table 3.1)

D_H = Empirical relationship given by equation (3.10) for Australia.

G = Daily aberage insolation per month, W/m^2

G_D = Diffused sky radiation, W/m^2

G_N = Direct normal radiation intensity at the surface of the earth

G_o = Extraterrestrial solar radiation, W/m^2 (Table 3.1)

G_{om} = Monthly average extraterrestrial solar radiation, W/m^2

G_R = Solar radiation reflected from surrounding surfaces, W/m^2

G_T = Total irradiation on a terrestrial surface, W/m^2

K_T = Cloudiness index

L = Latitude, degrees

R = Relationship defined by equation (3.9)

R_D = Tilt factor for north-facing surfaces

w_s = The sunset hour angle, radians

α = Incidence angle referred to a line normal to the surface, degrees (Figure 3.1)

β = Surface azimuth angle referred to a line normal to the intersection between the surface and the horizontal plane, degrees (Figure 3.1)

ϕ = Solar azimuth angle, degrees (Figure 3.1)

ψ = Surface tilt angle, degrees (Figure 3.1)

ρ_g = Solar reflectance of foreground surfaces (Table 3.2)

θ = Solar altitude angle, degrees (Figure 3.1)

APPENDIX 1

SOURCES OF SOLAR DATA FOR AUSTRALIA

The Secretary,
Department of Environment, Housing and Community Development,
Lombard House, Allaba Street (P.O. Box 1890)
Canberra City,   A.C.T., 2601
Australia

C.S.I.R.O.
Division of Atmospheric Physics,
P.O.   Box 77,   Mordialloc, Victoria, 3195
Australia

C.S.I.R.O.
Division of Mechanical Engineering,
P.O. Box 26,   Highett, Victoria,   3190
Australia

Bureau of Meteorology,
162 Goulburn Street,   Sydney,   2000
Australia

George O.G. Lof et al.,
World Distribution of Solar Radiation,
Engineering Experiment Station Report No.21,
University of Wisconsin,   Madison, WI   53706
U.S.A.

World Meteorological Organisation,
Monthly Averages for Global Insolation,
W.M.O.,   Geneva,   Switzerland

REFERENCES

1. American Society of Heating, Refrigerating and Air Conditioning Engineers, Inc., "ASHRAE Handbook: 1977 Fundamentals", ASHRAE, New York.

2. Paltridge, G.W. and Proctor, D., "Monthly Solar Radiation Statistics for Australia", vol. 18, Solar Energy (1976) pp.235-243.

3. Morrison, G.L., Sapsford, C.M. and Litvak, A., "Solar Insolation Data for Sydney", University of N.S.W., School of Mechanical and Industrial Engineering, Report No.1979/FMT/2.

4. Liu, B.Y.H. and Jordan, R.C., "A Rational Procedure for Predicting the Long Term Average Performance of Flat Plate Solar Energy Collectors", vol. 4, Solar Energy (1960) pp. 1-19.

5. Australian Department of Housing and Construction, "Air Conditioning Systems Design Manual" (1974).

6. Carrier Air Conditioning Co., "Handbook of Air Conditioning Systems Design", McGraw-Hill Book Co. (1965).

CHAPTER 4

Solar Collectors

INTRODUCTION

The primary function of solar collectors is to gather heat. Since they are usually the most visible and the most expensive part of the solar heating system, design details and material selection for solar collectors receive considerable attention (e.g. [1, 2]). For nearly a century the most popular device for the collection of solar radiation for domestic hot water and for home heating has been the flat-plate collector. Figure 4.1 shows examples of flat-plate collectors that are used to heat water and air. The basic flat-plate collector contains an absorber and flow passages for the coolant. Often, the absorber is just a thin metal sheet coated with a material that absorbs the solar energy. A coolant flowing in the passages collects the heat from the absorber and transports it to the thermal storage system. If the collection temperature is to be significantly above the ambient temperature, one or more layers of a cover material (usually either glass or a clear plastic) are placed above the plate. The main function of the cover material is to reduce the convective heat losses from the absorber without significantly reducing the intensity of the solar radiation that reaches the plate. These covers are transparent to the incoming short-wavelength solar radiation, but may absorb the low-temperature (long-wavelength) radiation emitted by the absorber and some of that heat will return to the absorber. This effect is called the "greenhouse effect".

In active systems, an absorber with a high thermal conductivity is desirable and the plates are usually either copper or aluminium, although steel and some plastics are occasionally used. The absorber plate is usually coated with a material which has a high absorptivity, such as a flat-black paint. More expensive selective coatings are available which have low emissivities for long-wavelength radiation which will reduce the radiant heat loss from the absorber.

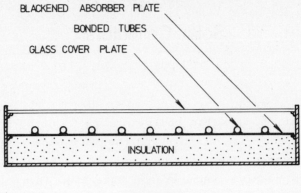

LIQUID TYPE

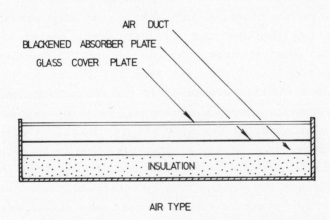

AIR TYPE

Fig. 4.1.   Examples of liquid and air cooled solar collectors.

Other components of the flat-plate collector might include
insulation to reduce heat losses from the back and sides of the
collector, a sealed case to keep moisture and dirt off the absorber
surfaces, and a special mounting for the collector.

Collector designs vary considerably depending on the application.
For example, collectors used to heat swimming pools only require a
small temperature rise through the collector (usually less than
$10^{\circ}C$) and the collector temperature is usually close to the ambient
air temperature.   In fact, the operating temperature of these

collectors is often below the ambient air temperature and the air actually helps heat the water in the collector. Consequently, the collector cover is often not required. In this case plastics can be used for the absorbing material and the cost of the collector can be reduced significantly.

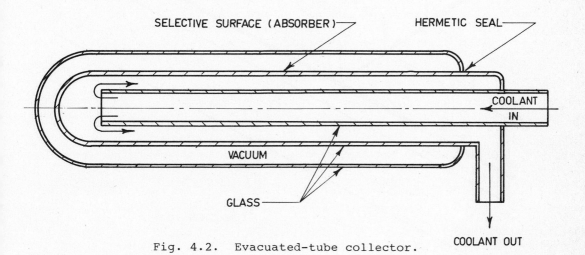

Fig. 4.2. Evacuated-tube collector.

Another type of collector is the evacuated-tube collector of the type shown in Figure 4.2. In this type, evacuation of the air between the absorber and the outer glass envelope eliminates convection losses from the absorber. Also, selective surfaces are used on the absorber to reduce the radiation losses, so this type of collector is particularly effective for high-temperature applications, such as solar air conditioning.

GENERAL PRINCIPLES AND PERFORMANCE

Energy Balance

Figure 4.3 shows the different modes of heat transfer that occur in a solar collector under steady-state conditions. The useful energy available from a solar collector (q_u) is calculated from an energy balance on this system [3].

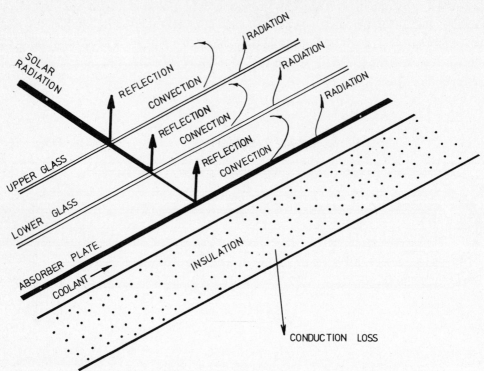

Fig. 4.3.   Heat exchange in a solar collector.

USEFUL ENERGY DELIVERED   = ENERGY ABSORBED − ENERGY LOST

or

$$q_u = G_T\ \tau_{su}\ \alpha_{su}\ A_c - U_L\ A_c\ (T_p - T_a) \tag{4.1}$$

where

G_T is the insolation on the collector (W/m^2).

τ_{su} is the solar transmissivity of the cover(s), i.e.
       the fraction of the solar radiation that reaches
       the absorber (dimensionless).

α_{su} is the solar absorptivity of the absorber, i.e. the
       fraction of the solar energy reaching the absorber that
       is actually absorbed (dimensionless).

A_c is the total collector area (m^2).

U_L is the overall heat-transfer coefficient between
        the absorber to the surroundings ($W/m^2\ ^\circ C$).

T_p is the average absorber plate temperature ($^\circ C$).

T_a is the ambient temperature of the atmosphere ($^\circ C$).

This equation relates the principal design factors that affect collector performance, specifically those involving the collection of solar radiation (i.e. the transmissivity and absorptivity) and those related to the control of heat losses.

Heat Losses

Equation (4.1) shows that the heat losses from the collector should be kept small. With reasonable care the insulation around the edges and on the bottom of the collector will effectively minimise the heat lost from those regions. These losses should only constitute 10-20% of the total heat lost from the collector. However, the heat lost from the top of the plate is far more difficult to reduce. If a collector plate without a glazing layer is used, the overall heat loss coefficient (U_L) would be so large that the heat lost to the environment would equal the solar energy absorbed by the collector when the temperature of the absorber plate is only 25°C above the ambient temperature. The 25°C temperature difference can only be achieved under stagnation conditions when the useful energy extracted from the collector (q_u) is zero. Consequently, except for applications where very small temperature differences are required (e.g. swimming-pool heating), it is necessary to reduce the heat losses from the top of the collector. This is usually achieved by placing one or more layers of glazing above the plate (Figure 4.3). Glazing does significantly reduce the amount of solar radiation absorbed by the collector; however, each glazing layer decreases the heat losses from the plate in two ways. Firstly, since stagnant air is an excellent insulator, the air trapped by the glazing layers can reduce the convection losses by a factor of 10 or more. Secondly, the low-temperature radiation emitted by the absorber will be absorbed by the glass and some of this energy will be retained as useful heat.

Stagnation Temperature

Equation (4.1) can be rearranged to show the factors that determine the average temperature of the absorber (T_p) under various operating conditions.

TABLE 4.1   Stagnation Temperatures for Different Types of Solar Collectors [3,4]

Collector Number	Absorber Material	Manufacturer and Remarks	Absorber Surface Coating	Transparent Covers		Stagnation Temperature* °C
				Number	Material	
1	Aluminium	NASA/Honeywell	Black Nickel	2	Glass	241
2	Aluminium	MSFC	Black Nickel	2	Tedlar	156
3	Aluminium	NASA/Honeywell	Black Paint	1	Glass	134
4	Aluminium	NASA/Honeywell (Mylar Honeycomb)	Black Paint	2	Glass	246
5	Aluminium	NASA/Honeywell	Black Paint	2	Glass	180
6	Aluminium	PPG	Black Paint	2	Glass	131
7**	Glass	Owens (Evacuated Tube)	Selective Surface	1	Glass	336
8**	Steel	Solaron (Data furnished by manufacturer) heat transfer fluid is air	Black Paint	2	Glass	180

*Values are calculated assuming that incident solar radiation, G_T, is 946 W/m^2 and that ambient temperature, T_a, is 21°C.

**With the exception of solar collectors number 7 and 8, the absorbers are the tubes-in-plate types.

$$T_p = T_a - [(q_u/A_c) - G_T \tau_{su} \alpha_{su}]/U_L \qquad (4.2)$$

Expressed in this form, the equation shows that the average plate temperature will increase as the useful energy (q_u) removed from the collector is decreased.   Under stagnation conditions (i.e. when no coolant is circulated in the collector) q_u is zero and the plate temperature reaches its maximum value (T_s),

$$T_s \text{ (STAGNATION TEMPERATURE)} = T_a + G_T \tau_{su} \alpha_{su}/U_L \qquad (4.3)$$

If overheat protection is not provided, then it is inevitable that some collectors will experience stagnation conditions for varying periods during their operating life [4].   These stagnation temperatures can be reached during installation (when the collector is exposed to the sun before it is connected to the rest of the system) and during periods when the system is shut down (e.g. during power failures, or when the owners are away on holiday, or when the system is being overhauled, or after the house has been sold but the new owners have not yet taken possession).

Table 4.1 shows the stagnation temperature for several different types of collector.   At these temperatures, most plastics and many non-metals degrade and the binder used in the insulation as well as some of the components used in paints may volatise.   Since the glass cover is the coolest surface inside the collector, these vapours will usually condense on the cover and reduce its transmissivity.   In addition, if the collector has been stagnating and a relatively cool fluid is suddenly circulated through the collector thermal shock will occur which can cause the absorber plate to buckle and to separate from the tubes [4].

EXAMPLE 1   Solar Collectors

A solar collector with characteristics similar to those of the single glazed collector shown in Figure 4.6 (i.e. $\tau_{su} \alpha_{su} = 0.80$, $U_L = 8.0$) is used in a domestic hot water system.   The collector faces due north on a roof in Sydney (34OS latitude) and is tilted at

50° to the horizontal (ψ).   For 11 a.m. Australian Eastern Daylight Saving Time on December 21st, find

a)   The insolation (G_T) on the collector.

b)   The useful energy collected if the average the plate temperature (T_p) is $65^\circ C$ and the ambient temperature (T_a) is $22^\circ C$.

c)   The "threshold" radiation level for this case (i.e. the lowest value of G_T for which heat will be collected).

d)   The collector stagnation temperature (T_s) under these conditions.

Neglect both the reflected and the diffuse components of G_T.

Solution:

a)   From Examples 1 and 2 in Chapter 3, the solar time is 10.093 hours and the solar angles are

$$\theta = 62.02^\circ \qquad \text{and} \qquad \phi = 72.79^\circ$$

From equation (3.1a)

$$G_N \; = \; A \, \exp[\, - \, B/\sin \theta \,]$$

where A and B are constants obtained from Table 3.1.   Thus,

$$G_N \; = \; 1.23 \, \exp[\, - \, 0.142/\sin 62.02^\circ]$$
$$G_N \; = \; 1.047 \; kW/m^2$$

Also, from equation (3.1)

$$G_T \; = \; G_N \, \cos \alpha$$

where, α is the angle between the sun's rays and a normal to the plate.   From equation (3.1d),

$$\cos \alpha \; = \; \cos \theta \; \cos \beta \; \sin \psi \; + \; \sin \theta \; \cos \psi$$

$$\cos \alpha = \cos 62.02^{\circ} \cos 72.79^{\circ} \sin 50^{\circ} + \sin 62.02^{\circ} \cos 50^{\circ}$$

$$= 0.6740$$

and

$$\alpha = 47.62^{\circ}$$

so,

$$G_T = 1.047 \cos (47.62^{\circ})$$
$$G_T = 0.7057 \text{ kW/m}^2$$

b)   The useful energy collected (q_u) is obtained from,

$$q_u = G_T \tau_{su} \alpha_{su} - U_L (T_p - T_a)$$
$$q_u = 705.7 (0.80) - 8.0 (65 - 22)$$
$$q_u = 220.5 \text{ W/m}^2$$

c)   The threshold radiation level occurs when $q_u = 0$

$$0 = G_T (0.8) - 8.0 (65 - 22)$$
$$G_T = 430 \text{ W/m}^2$$

d)   From equation (4.3),

$$T_s = T_a + G_T \tau_{su} \alpha_{su}/U_L$$
$$= 22 + 705.7(0.8)/8.0$$
$$T_s = 92.6^{\circ}C$$

COLLECTOR RELIABILITY AND MATERIALS PERFORMANCE

RELEVANT FACTORS

Factors that affect the reliability and performance of solar collectors can be divided into long-term and short-term effects. The short-term effects are usually caused by design faults that become evident within the first few years of operation, while the long-term effects usually appear only after five years or more of operation and generally result in a slow degradation in collector performance.   Some long-term factors might be:

1.   Corrosion or oxidation of the case or of the absorber surface.
2.   Slow scaling of the coolant passages caused by impurities in the coolant.
3.   Low-cycle fatigue at the bond between the coolant passages and the absorber plate caused by thermally induced cyclic stresses.

Recent studies of 66 solar demonstration projects in the U.S. have identified several major causes of short-term failures in solar home-heating and cooling systems.   Data analysed at the Argonne National Laboratories [5 - 7] showed that the parts of these systems most likely to experience short-term failures were:

1.   The collectors themselves [5].
2.   The interconnections between the collectors and the manifolds [6].   These were mainly due to hose-clamp leakage, material incompatibility, workmanship and thermal expansion.
3.   Parts of the system susceptible to freezing [5].
4.   The control systems [7].

Although all 66 systems had only operated for a few years, 25 experienced collector problems.   These problems included cover-plate breakage (usually caused by thermal stresses), leaks, weathering, freezing and mechanical failures.   In addition, in at least one system, the solder bond between the coolant tube and the

absorber plate had fractured (Figure 1.10)*.   Other mechanical
difficulties were related to design and included the retaining clip
and collector bracket design, deterioration of sealant materials and
failure to allow for differential thermal expansion within the
collector assembly.

Glazing Materials

Transmissivity and absorptivity

When radiation strikes the interface between two materials (e.g. one
of the air-glass interfaces in Figure 4.3) some of the radiation is
reflected.   The proportion of radiation reflected is a function of
the wavelength of the radiation, the properties of the materials,
the angle of incidence at the interface, the surface finish and many
other factors.   In addition to the losses due to reflection, some
of the radiation that crosses the interface is absorbed in the
glass.   The main factor that determines the absorption of solar
energy in glass is its iron oxide ($Fe_2 O_3$) content.   Glasses such
as "white water" glass with iron-oxide content well under 0.15% are
preferred for use in solar collectors.   Unfortunately, low-iron
glass is relatively expensive and was not used in most of the solar
collectors that are presently installed.

Table 4.2 shows the approximate transmissivities of several
materials that are commonly used as covers on solar collectors.
These values are only approximate since the cover thickness, the
angle of incidence of the sun's rays, the surface finish, dirt
accumulation, etc., all have an effect on the transmissivity of
cover materials.

A solar collector with multiple layers of glass is quite complex to
analyse since some solar radiation is reflected at each interface
and some of this reflected radiation is re-reflected at other
interfaces (Figure 4.3).   In addition, the radiation is attenuated

*Fracture of the tube-to-plate bond is also a major cause of
 long-term degradation in older solar domestic hot water systems.

TABLE 4.2   Solar Transmissivity (τ_{su}) for Various Cover Materials [3]

Material	Transmissivity (τ_{su})
Low-iron glass	0.91
Window glass	0.85

Polymethyl Methacrylate (Acrylic)
    Acrylite
    Lucite
    Plexiglass
    Perspex
	0.80

Polycarbonate
    Lexan
    Merlon
	0.84

Polyethylene Terephthalate (Polyester) Mylar	0.84
Polyvinyl Fluoride Tedlar	0.93
Polyamide Kapton	0.80
Polyethylene	0.86

Fluorinated Ethylene Propylene
    (Fluorocarbon)
    FEP Teflon
	0.96

Fibreglass-reinforced Polyester Kalwall	0.87

Fibreglass-reinforced, Acrylic-
    fortified Polyester with
    Polyvinyl fluoride weather surface
 	
    Tedlar-clad Filon
	0.86

during its passage through each glazing layer. Consequently, the analysis of solar collectors with multiple glazing layers is usually done with the aid of a computer [3]. For design calculations, it is reasonable to assume that ordinary window glass will reflect about 8% of the solar radiation that arrives perpendicular to its surface. Two sheets of glass with an air gap between will reflect about 15%. In addition, low-iron glass will absorb about 2% of the incident radiation for a 3-mm thick sheet and normal window glass will absorb about 5% per sheet. If the angle of incidence between the radiation and the collector varies throughout the day as well as seasonally, the transmitted energy will be even lower. With low-iron glass the average absorption-transmission coefficient ($\tau_{su} \alpha_{su}$) would be about 0.80 for a single-glazed collector and about 0.70 for one with double glazing. Some plastics are used for the glazing material and their transmissivities are usually somewhat higher than that of glass.

Metallic films formed by vapour deposition on the glazing surface and surface etching techniques have been proposed that reduce the reflectivity of glass. Although the performance of the solar collector can be substantially improved by these techniques, they are expensive and have not yet been proved to be cost effective.

Materials assessment

If a solar collector is installed on the roof of a home, the homeowner will probably be disappointed if the useful life of the collector is less than 20 years. In fact, experience with solar domestic hot-water heaters in Florida shows that homeowners expect the solar collectors to operate with a minimum of maintenance for the life of the roof itself. Toughened, low-iron glass is the only cover material that can be expected to last over 20 years without significant degradation. It is scratch resistant, weather resistant and has good thermal stability. Its thermal expansion coefficient is compatible with metals, so thermally-induced stresses are less of a problem than with most plastics. However, glass is relatively heavy, it is brittle and significantly more expensive than some plastics.

Cheng [8] reported the results of materials-assessment studies conducted at solar demonstration sites throughout the U.S.   His report on glazing materials indicated:

1. Untempered glass is susceptible to shattering during hailstorms. Tempered, low-iron glass used in the demonstration programs had good shatter resistance and also good transmission in the range of 0.86 (or 0.90 if the glass is acid etched).

2. Dirt and grime deposited over time was found to be one of the most serious threats to the durability of glass; however, the problem was found to be no worse for etched than for unetched glass.   In one experiment in an urban environment, dirt and grime deposited on the outside surface of the glass reduced its transmissivity by 4% in six months.   Washing with a mild detergent solution easily restored the glass to its original condition.

3. For some collector designs, outgassing of materials used inside the collector (e.g. insulation) deposited on the underside of the glass.

Relative to glass, plastics weigh less and have better fracture resistance for the same thickness.   However, most plastics are too fragile and their weather resistance too low to be used as cover materials.   In less than one year of use, most sheets and films will yellow, become brittle and deteriorate.   Also, mechanical stress and wind-flutter will reduce their useful life.   Plastics do not have the high temperature tolerance of glass and will sag significantly when subjected to high humidity and to temperature differentials.   In addition, plastics are less scratch resistant and they are usually transparent to long wave-length radiation. Two plastic sheetings that are occasionally used as covers for solar collectors are,

1.   Polycarbonate plastics (e.g. Lexan):   This plastic is shatter resistant and is not as heavy or breakable as glass, but it has a high thermal expansion coefficient and creeps at high stresses and temperatures.   For these reasons, excessive warpage and

stress cracking has occurred in several polycarbonate collector covers. Also, although polycarbonate has fairly good transmissivity initially (between 0.845 and 0.88), in solar applications yellowing due to surface weathering and ultraviolet degradation can cause this to drop by 5% after a few years service [1,2,8].

2.   Acrylics (e.g. Plexiglass, Perspex):   As collector covers, these plastics have good transmission at solar wavelength, but they will also transmit most of the long-wavelength radiation emitted by the absorber plate.   In addition, since acrylics have a high thermal expansion coefficient and creep signficantly at temperatures above $75^{\circ}C$, they can develop high stresses and distort significantly under stagnation conditions [1,2,8].

Plastic films are attractive due to their low cost, but need supports to stretch and hold them in position. Plastics generally degrade when exposed to radiation; however, ultraviolet durability may not be required if the plastic is used as an inner cover. Cheng listed the following plastic films in order of preference according to their transmission and durability:

1.   TEFLON FEP (Fluorocarbon, DuPont):   This film has a maximum long-service temperature of $200^{\circ}C$, and has average to good weatherability based on tests in Arizona, Florida and Pennsylvania; however, it is more expensive than toughened glass [1].   Its useful life is probably about 20 years.

2.   TEDLAR PVF (Fluorocarbon, DuPont):   This fluorocarbon film is significantly cheaper than glass and is probably the most common of all the plastic cover materials.   It has a maximum long-term service temperature of $105^{\circ}C$ and was rated as having good to excellent weatherability after ten years exposure in Florida [1].   However, its transmissivity did decrease by 5% in five years [2].   Its useful life is probably less than ten years.

3.   SUNLITE PREMIUM (FRP, Kalwall):   This fibreglass reinforced plastic is also less expensive than glass, but prolonged use above $93^{\circ}C$ will reduce its transmissivity by over 10% [2].

4. MYLAR (Polyester, DuPont):- This film is the least expensive of all the cover materials discussed and has a maximum long-service temperature below 150°C. However, it degrades rapidly when exposed to UV radiation [5] and has a useful life of less than 4 years [2].

Solar Absorber

The solar absorptivity of the collector plate, a_{su}, is closely related to its optical properties. Dark coloured surfaces have a high absorptivity for the visible portion of the spectrum and are usually good absorbers in the infrared portion of the solar spectrum. Carbon black, many metal oxides and black paints absorb more than 94% of the incident solar radiation. Heat-resistant black paints are the most commonly used absorber coatings for solar collectors. These paints are usually sprayed on the collector and then cured at high temperatures to drive off the solvents and other volatile materials. These surfaces must be able to withstand long exposure to temperatures above 150°C without appreciable deterioration or outgassing since volatile gases released by the paint (or from the insulation) can coat the glazing material and reduce its transmissivity significantly. Temperature cycling and UV radiation also cause many types of paint to fail after a few years inside a solar collector [9].

Most surfaces that are good absorbers of solar radiation are also good emitters of long-wavelength (low temperature) radiation. However, there are a few surfaces, called selective surfaces, that are good absorbers of solar radiation, but are poor emitters of low-temperature radiation. The properties of two promising selective surfaces are shown in Figure 4.4. They consist of a thick black-metallic oxide (electro-deposited black nickel or black chrome) on a bright nickel substrate. The oxide is a good absorber of solar radiation but is so thin that it is essentially transparent to low-temperature radiation. Consequently, the amount of heat lost by radiation from the collector is determined by the low emissivity of the bright-metal substrate. These selective surfaces have solar absorptivities above 0.90 and thermal emissivities below 0.1 at normal operating temperatures.

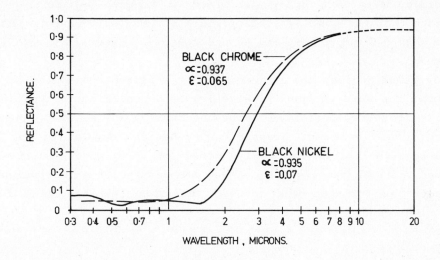

Fig. 4.4. Reflective properties of two selective surfaces.

Since selective surfaces are electroplated on the absorber plate, damaged surfaces, or surfaces that have deteriorated in use, cannot be repaired at the site but would probably have to be removed from the system and returned to the manufacturer for repair. Consequently, to be cost effective, selective surfaces must offer long-term durability when exposed to weathering, humidity, sunlight, thermal cycling and high stagnation temperatures. Data in Table 4.1 for collectors 1 and 5 indicate that the stagnation temperature of a collector can increase by more than $50^{\circ}C$ if a selective surface is used. Tabor [11] indicated that, since selective surfaces are so thin (on the order of 0.1 micron or less), they are unlikely to effectively protect the base material against atmospheric corrosion or thermal oxidation. Therefore, both the coating and the base material must be chemically and thermally stable. Few low-cost materials can withstand condensation or even humidity for long periods [11] and, as yet, no selective surface has been shown to have a 20-year life in a flat-plate solar collector application. However, work at Los Alamos indicates that this goal may be reached soon and, of the surfaces presently available, black chrome probably has the best chance of achieving this goal [1,11]. Results from the materials assessment study performed at Argonne Laboratories [4] indicate that the coating life of black chrome could be more than 30 years when 0.5-mil nickel and an undercoat are used.

At low temperatures, the performance of a flat-plate collector will be about the same whether a selective or a non-selective coating is used.   If the collector is to be used solely for heating, then the lower cost, ease of repair and better durability of the non-selective coating probably gives it the edge at present. However, if both heating and cooling are desired, the improvement in collector performance will offset the high initial cost of the selective coating [9].

Insulating Materials

Cheng [8] also studied the durability of various insulating materials and reported the following:

1.   Mineral Wool: Mineral wool tends to settle under its own weight and leave an air gap between it and the absorber panel, increasing the heat losses from the bottom of the absorber plate and decreasing the collector efficiency with time. Humidity cycling will cause it to degrade, so the collector housing must be hermetically sealed. In addition, loose insulation complicates repair and maintenance.

2.   Polyurethane Foam: Exposed polyurethane foam cannot be recommended since field experience showed exposure to ultraviolet radiation caused degradation of the material followed by swelling.

3.   Fibreglass: Semi-rigid fibreglass board with hardly any binder, which is suitable for high-temperature applications, is recommended. The insulation should be preheated above $177^{O}C$ to drive off the volatile binder and any residual moisture just before the glass cover is put in place.

COLLECTOR EFFICIENCY

MATHEMATICAL MODEL

The collector efficiency (η) is the ratio of the energy absorbed by the working fluid (the useful energy) to the solar energy incident upon the collector (the insolation). That is

$$\eta = q_u/[G_T \ A_c] \qquad\qquad (4.4)$$

or, from equation (4.1)

$$\eta = \tau_{su} \ \alpha_{su} - U_L \ (T_p - T_a)/G_T \qquad\qquad (4.5)$$

With the exception of the average plate temperature (T_p) all of the parameters in equation (4.5) can readily be measured or obtained from available data. The average plate temperature is difficult to determine as it is greatly influenced by the inlet temperature of the coolant and, to a lesser extent, by other operating conditions. Therefore, to simplify the analysis, equation (4.5) is usually modified by the substitution of the inlet fluid temperature (T_i) for the average plate temperature and a correction factor called the heat recovery factor (F_R) is applied to the resulting equation.

$$\eta = F_R \ \tau_{su} \ \alpha_{su} - F_R \ U_L \ (T_i - T_a)/G_T \qquad\qquad (4.6)$$

The heat recovery factor has a value between 0 and 1 and is determined experimentally for each collector.

For a specific coolant flow rate, the values of F_R, τ_{su}, α_{su} and U_L are practically independent of the system temperature and of the solar insolation for any given collector. Consequently, the collector efficiency is a linear function of $(T_i - T_a)/G_T$ and can be approximated by a straight line on an η vs $(T_i - T_a)/G_T$ plot. Such a plot is shown in Figure 4.5 for two flat-plate collectors. Both of these collectors are tube-in-sheet collectors with flat-black paint used for the absorber surface. The only significant difference between them is that one has a single glass cover and the other is double glazed. This graph shows the collector efficiency

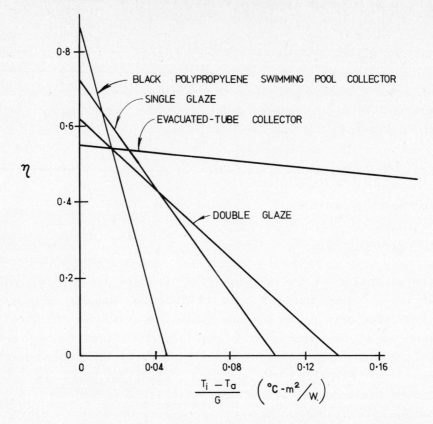

Fig. 4.5.   Effect of double glazing on collector efficiency.

under steady-state operating conditions.   The straight-line
intersects the vertical axis at $T_i = T_a$ and

$$\text{intersect} = F_R \, \tau_{su} \alpha_{su} \tag{4.7a}$$

and the slope of the line gives

$$\text{slope} = - F_R \, U_L \tag{4.7b}$$

Approximate parameters for the six basic collector types that are
available in Australia are given in Table 4.3.   These performance
characteristics were not obtained under the same test conditions or
even at the same test facility and most of the collectors were new
when tested (only collector number 5 had experienced stagnation

conditions).   Consequently, the data probably do not accurately represent the long-term performance of these collectors (see Chapter 5 for long-term performance of collectors).   These data can be used for preliminary analyses to size systems and to predict the relative initial performance of the various system types.   However, before any system is finally selected, accurate cost and long-term performance data should be obtained from the various manufacturers.

TABLE   4.3   Approximate Performance Parameters for the Six Types of Flat-Plate Collectors Marketed in Australia

No.	Absorber Plate	Absorber Surface	Cover Material	Insulation	$F_R \tau_{su} \alpha_{su}$	$F_R U_L$ W/m^2 $^\circ C$
1	Aluminium fin, thermal grease and copper tube	Black paint	Glass (1 layer)	50-mm fibre glass	0.66	6.3
2	Copper fin, solder and copper tube	Black paint	Glass (1 layer)	50-mm fibre glass	0.72	7.1
3	Copper fin, solder and copper tube	Copper oxide selective surface	Glass (1 layer)	50-mm fiber glass	0.68	4.7
4	Copper fin, solder and copper tube	Black chrome selective surface	Glass (1 layer)	50-mm fibre glass	0.80	5.7
5	Expanded stainless steel	Black chrome selective surface	Glass (1 layer)	50-mm fibre glass	0.77	4.2
6	Plastic (swimming pool type)	Black	none	none	0.73	22.2

EXAMPLE 2 :   Collector Operation

The table below gives the hourly values for the insolation (G_T),
ambient temperature (T_a), fluid inlet temperature (T_i) and the
average absorber plate temperature (T_p) measured on the 21st of
December.   For the single-glazed collector in Figure 4.4 (collector
2 in Table 4.3), find:

    a)   The "useful" energy collected that day (q_u)
    b)   The collector efficiency (η) for that day
    c)   The "effective" value of $(T_i - T_a)/G_T$ for that day

$$T_p = 65^\circ C$$

Environmental Data

Time	G_T W/m^2	T_i $^\circ C$	T_a $^\circ C$
7 am	27.2	60.0	18.8
8 am	178.1	60.0	19.1
9 am	584.6	60.0	19.4
10 am	853.0	60.0	20.9
11 am	983.7	60.0	22.0
Noon	1033.5	60.0	22.6
1 pm	1086.2	60.0	23.2
2 pm	1027.3	60.0	23.7
3 pm	943.4	60.0	24.4
4 pm	822.1	60.0	24.2
5 pm	538.9	60.0	22.9
6 pm	154.2	60.0	22.0
7 pm	32.8	60.0	20.5

Solution

a)   The net heat gain must be calculated for each hour of operation.
    This can be accomplished by either using the basic collector
    parameters that were given in Example 1 (Method 1), or from the
    measured parameters for this collector given in Table 4.3
    (Method 2).

Method 1

From Example 1,

$$\tau_{su}\alpha_{su} = 0.8, \quad U_L = 8.0 \text{ W/m}^2 \text{ }^\circ C$$
and
$$q_u = \tau_{su}\alpha_{su} G_T - U_L (T_p - T_a) N$$

where N is the number of hours during the day that G_T is above the threshold value of 372.3 W/m^2.   In this case, N is 9 hours, G_T (total) is 7872.7 W/m^2 and T_a for those 9 hours is 22.6°C. Thus,

$$q_u \text{ (total)} = 0.8 (7872.7) - 8.0(65 - 22.6)9$$
$$= 3.245 \text{ kWh/m}^2 \quad (11.7 \text{ MJ/m}^2 \text{) for that day.}$$

Method 2

1. Calculate the hourly values of $(T_i - T_a)/G_T$ and enter these values in the table below.

2. Obtain η from Figure 4.4 for each hour and enter these values in the table.

3. Calculate useful energy for each hour of operation from
    $$q_u = \eta G_T$$
    and enter these values in the table.

4. Sum the hourly values of q_u in the last column in the table to give the total useful energy collected for that day, which is,

    $$q_u \text{(total)} = 3.262 \text{ kWh/m}^2 \quad (11.7 \text{ MJ/m}^2) \text{ for that day.}$$

b) Calculate the "effective" collector efficiency from

    $$\eta \text{(eff.)} = \text{Total heat collected that day/Total insolation}$$

    The totals for the daily heat collected and for the insolation are given in the table below.   Thus,

    $$\eta \text{(eff.)} = 3262.8/7872.7 = 0.414$$

c) Figure 4.4 shows that this efficiency corresponds to an average value of $(T_i - T_a)/G_T$ of 0.043.   This value is close to the one given for domestic water heaters in Table 4.4.

The tabulated results for Example 2 are,

Time	G_T W/m^2	T_i $°C$	T_a $°C$	$(T_i-T_a)/G_T$ $(m^2 °C/W)$	η	ηG_T W/m^2
7 am	27.2	60.0	18.8	1.5147	0.00	0.0
8 am	178.1	60.0	19.1	0.2296	0.00	0.0
9 am	584.6	60.0	19.4	0.06544	0.26	152.0
10 am	853.0	60.0	20.9	0.04584	0.39	332.7
11 am	983.7	60.0	22.0	0.03863	0.45	442.7
Noon	1033.5	60.0	22.6	0.03619	0.46	460.6
1 pm	1086.2	60.0	23.2	0.03388	0.48	521.4
2 pm	1027.3	60.0	23.7	0.03534	0.47	483.1
3 pm	943.4	60.0	24.4	0.03774	0.46	434.0
4 pm	822.1	60.0	24.2	0.04355	0.40	326.6
5 pm	538.9	60.0	22.9	0.06884	0.25	109.7
6 pm	154.2	60.0	22.0	0.2464	0.00	0.0
7 pm	32.8	60.0	20.5	1.2043	0.00	0.0
TOTALS	7872.7*		22.6*(average)			3262.8*

* Excludes terms for those hours when G_T is below the threshold value of 372.3 W/m^2 .

The heat recovery factor (F_R) is a function of the absorber plate fin efficiency (η_F), the collector area (A_c), the overall collector heat-loss coefficient (U_L), the mass flow rate of the coolant (m_c), the heat capacity of the coolant (c_c), the area of the coolant flow passages measured parallel to the absorber plate (A_f) and the overall heat-transfer coefficient between the coolant and the absorber plate (U_f).   This relationship is,

$$F_R = [(1/\eta_F) + (U_L A_c)(1/(U_f A_f) + 1/(2 m_c c_c))]^{-1} \quad (4.8)$$

This equation is quite useful as a design tool, but its application is beyond the scope of this book.   However, two examples are given in Appendix 1 which demonstrate how all of these equations might be used to help evaluate flat-plate collectors for particular applications.   One of the examples is for a single-glazed collector and the other for a black plastic swimming-pool collector.

APPLICATION OF THE MODEL

ABSORBER PLATE PARAMETERS

To select the "best" collector for a particular application, the designer must know the useful energy that will be delivered per dollar of installed cost for each of the collectors that are being considered.   Unfortunately, reliable long-term efficiency data are often not available for these collectors and the designer must rely on his past experience and analytical models to estimate the relative performance of the different types of collectors.   To demonstrate the use of the analytical model described in the last section, the model will be used to predict the effect of variations of tube spacing, absorber material and absorber thickness on the efficiency of a liquid-cooled, flat-plate collector.

EXAMPLE 3 : Absorber Design

The only major differences between several tube-in-plate solar collectors that are being considered for a particular application are the tube spacing, the absorber material and the absorber thickness.   Find the relationship between these parameters and their effect on the collector efficiency.

Solution:
The relationship between these parameters and the collector efficiency is developed in Appendix 2.   There it is shown that

$$\eta = \tau_{su}\alpha_{su} + U_L(T_o - T_a)/G_T - U_L(q_s - q_L)L^2/(3ktG_T) \quad (4.9)$$

Since the only factors on the right-hand side of this equation that differ between the collectors are the tube spacing (L), the absorber plate thickness (t) and its thermal conductivity (k), the efficiency of this group of collectors can be expressed as

$$\eta = C_7 - C_8L^2/(kt)$$

where C_7 and C_8 are constants.

Similarly, equations (4.2) and (4.9) give

$$q_u = G_T r_{su} \alpha_{su} + U_L (T_o - T_a) - U_L L^2 (q_s - q_L)/(3kt) \quad (4.10)$$

or

$$q_u = C_9 - C_{10} L^2 /(kt) \quad\quad\quad (4.11)$$

This equation indicates that the efficiency decreases with the square of the tube spacing and is inversely proportional to both the thermal conductivity and the thickness of the absorber plate. Consequently, this expression is an approximate relationship between these design parameters and the useful energy delivered per square metre of collector area. Thus, the collector with the smallest value of L^2/kt, delivers the most useful energy per square metre.

EFFICIENCY OF A FLAT-PLATE COLLECTOR

The steady state collector efficiency does not give an accurate prediction of the long-term operating efficiency of a collector since the fluid and ambient temperature as well as the insolation vary continuously during the day. An average daily efficiency for the collector is more useful for that purpose. However, these plots can be used to compare different collectors, each of which uses the same coolant. For example, the data in Figure 4.5 were obtained from two collectors of identical size and construction except that one collector was single glazed and the other double-glazed. The figure shows that the double-glazed collector is less efficient than both the single-glazed and the unglazed collector at temperatures close to the ambient, but more efficient at higher fluid temperatures. Consequently, for low-temperature applications, such as swimming-pool heaters where the required water temperature rise is less than $10^{o}C$, the unglazed collector is more efficient than either the single or double-glazed collector. Also, since a double-glazed collector might be 10% more expensive than the same collector with single glazing (and unglazed collectors are even less expensive), the actual application must be considered carefully before the appropriate collector can be selected. The following

table gives values of $(T_i - T_a)/G_T$ that are typical of several
applications [9].

TABLE 4.4   Typical Values of $(T_i - T_a)/G_T$

Application	$(T_i - T_a)/G_T$ m^2 $^{\circ}C/W$
Swimming pool	0.01
Reverse cycle air conditioner	0.04
Domestic hot water	0.05
Direct space heating	0.07

To select a collector for a specific application, the efficiency of
the various collectors being considered should be obtained for the
appropriate value of $(T_i - T_a)/G_T$.   These efficiencies should be
divided by their cost per square metre. If all other factors are
equal, the highest value of this efficiency-to-cost ratio will be
the best buy [9].   In practice, this approach should only be used
to determine which of the many available collectors are unsuitable
for a particular application.   Other factors (such as durability,
warranty offered, aesthetics, availability, etc.) must be considered
before the final choice can be made.

Comparison of Liquid- and Air-Heating Systems

Collector efficiency curves, such as those shown in Figure 4.5, can
be used to compare different liquid-heating collectors.   Comparison
of liquid collectors and air collectors is more difficult since the
air-temperature rise in a single pass through a collector may often
exceed $25^{\circ}C$, whereas the temperature of the water in these systems
is rarely increased by more than $8^{\circ}C$ in one pass.   The following
example, taken from reference [3], shows how these two systems might
perform under similar operating conditions.

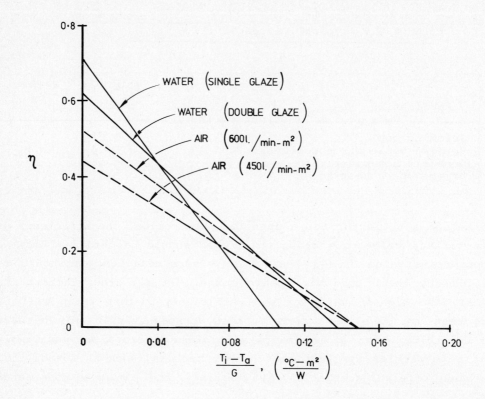

Fig. 4.6.   Typical efficiency curves for hot-air and hot-water collectors.

EXAMPLE 4   Collector Performance

Compare the performance of a water-cooled and an air-cooled collector under the following operating conditions:

G_T = 1000 W/m^2, T_a = 15°C
water: 0.8 kg/min (per m^2 of collector area); T_i = 45°C
air : 0.60 m^3/min (per m^2 of collector area); T_i = 20°C

Find the efficiencies and the coolant outlet temperatures for the two collectors.

Solution:

Water

$\qquad (T_i - T_a)/G_T \;\; = \;\; 0.03 \; m^2 \; ^oC/W$

Air

$\qquad (T_i - T_a)/G_T \;\; = \;\; 0.005 \; m^2 \; ^oC/W$

From Figure 4.6,

$\qquad$ water $\eta = \;\; 0.49 \qquad\qquad$ air $\eta = 0.47$

The coolant outlet temperature can be obtained from the useful energy absorbed, the mass flow rate and the inlet conditions.

$\qquad q_u \;\; = \;\; m_c c_c (T_O - T_i)$

and

$\qquad q_u \;\; = \;\;\;\; \eta \; G_T$

For the water-cooled collector,

$\qquad q_u \;\; = \;\; 0.49 \; (1000)$
$\qquad q_u \;\; = \;\; 490 \; W/m^2$

so,

$\;\; 490 \; W = \;\; 0.8 \; kg/min \; (4.17 \; kJ/kg^oC) \; (T_O - 45)^oC/(60 \; sec/min)$

and

$\qquad T_O \;\; = \;\; 53.8 \; ^oC$ (for water)

For the air-cooled collector,

$\qquad q_u \;\; = \;\;\;\; 1000 \; (0.47) \;\;\; = \;\;\; 470 \; W$
$\qquad 470 = 0.60 \; m^3/min(1.006 \; kJ/kg^oC)1.177 \; kg/m^3 (T_O - 20)^oC/(60 \; sec/min)$

and

$\qquad T_{out} \;\; = \;\;\;\; 59.7^oC$ (for air)

Consequently, under these operating conditions the water cooled collector delivers more useful energy; however, since the hot air is delivered at a higher temperature than the hot water, either collector might be preferred for different applications. Unfortunately, no easy method exists for comparing the performance of these two systems. However, a recent study of existing air and liquid systems indicated that the air systems collect more useful heat than water systems operated under similar conditions [3].

The blowers on air-heating systems consume large amounts of electricity and air systems also require expensive duct work to take the hot air from the collectors to the thermal-storage system. They are noisier than liquid systems and they need much more space for thermal storage than liquid systems (roughly three times as much volume as for water storage). The cost of all these factors should be considered when the system is selected.

There are very few commercially available solar air heaters. The reason for this may be that hot water has more uses than hot air, or it may be due to the long-established art of solar water heating, or simply because air-heating collectors are relatively easy for the home handyman to construct himself. These systems are inherently more reliable than water systems (one U.S. manufacturer guarantees his air collectors for 10 years) and air leaks are less catastrophic to the home owner than water leaks.

In summary, water systems have the advantage of a common heat transfer and storage medium and the storage volume is about one-third that of an equivalent air system. Liquid systems are more compatible with absorption air-conditioning systems than are hot-air systems. Liquid systems are also less noisy than air systems and require less electrical power to move the fluid. (Typically, the electrical requirements of a liquid system range from 6 to 8% of the useful solar energy delivered.) In addition, it is usually easier to install small water pipes than large air ducts in an existing structure. On the other hand, air systems have no corrosion, boiling or freezing problems and probably have lower maintenance costs than liquid systems.

APPENDIX 1

THEORETICAL PERFORMANCE OF TWO FLAT-PLATE COLLECTORS

EXAMPLE A :   Single Glazed Collector

For the collector in Example 1,

U_f = 1200 W/(m^2 $^{\circ}$C), A_c = 3 m^2, A_f = 0.5 m^2, m_c = 0.040 kg/s,
c_c = 4.19 kJ/(kg $^{\circ}$C), T_i = 60°C, T_a = 10°C, η_F = 0.99
G_T = 980 W/m^2, $\tau_{su}\alpha_{su}$ = 0.8, U_L = 8.0 W/(m^2 $^{\circ}$C)

a)   Calculate F_R for the collector.
b)   Calculate the heat transferred to the coolant (q),
     the outlet temperature (T_o) and the mean water
     temperature (T_m).
c)   If the coolant flow rate is reduced by one half,
     calculate the new vales for F_R, q and T_m.
d)   Find the coolant flow rate (m_c) for which T_m will be 85°C.
e)   Find the maximum possible coolant outlet temperature
     under these conditions.

Solution
a)
$$F_R = (1/0.99 + (8.0 \times 3)[1/(1200 \times 0.5) + 1/(2 \times 0.04 \times 4190)])^{-1}$$
$$F_R = 0.892$$

b)   First find the heat that is transferred to the coolant, i.e.
q (to coolant), then use

     q (to coolant)   =   q (stored in coolant)

To find T_o

     q (to coolant)   = [F_R τ_{su} $\alpha_{su}G_T$ - F_R U_L (T_i - T_a)] A_c
                       = [0.892(0.8)980 - 0.892(8)(60 - 10)] $\times$ 3
                       =   1027.6 W

$$q_{stored} = (m_c c_c)(T_i - T_o) \qquad\qquad (5.9)$$

$$T_o = T_i + q_{stored}/(m_c c_c)$$
$$= 60 + 1027.6/167.6 = 66.1^o C$$

Since $\eta_F = 0.99,$

$$T_p = T_m = (66.1 + 60)/2 = 63^o C$$

c) If the coolant flow rate is reduced by one-half but all other factors remain constant, then

$$F_R = [1/0.99 + (8x3)\ [1/(1200x0.5) + 1/(2x0.02x4190)]]^{-1}$$

$$F_R = 0.838$$

$$q(\text{to coolant}) = 1027.6 \times 0.838/0.892 = 965.4\ W$$

$$T_o = 60 + 965.4/83.8 = 71.5^o C$$
and
$$T_m = 65.8^o C$$

d) For these conditions, find the m_c for which

$$T_o = 85^o C$$

$$q(\text{stored}) = m_c c_c (T_o - T_i) = m_c(4190)(85-60)$$
$$q(\text{stored}) = 104,750\ m_c \quad (\text{in Watts})$$

$$q(\text{stored}) = q(\text{to coolant})$$
$$104,750\ m_c = [F_R\ \tau_{su}\ \alpha_{su} - U_L(T_i - T_a)]\ A_c$$

Substitution into equation (5.8) for F_R (which is also a function of m_c) gives

$$m_c = 0.007746\ kg/s$$
and

$$q = 104,750\ (0.007746) = 811.4\ W$$

e) Find maximum possible coolant outlet temperature

$$T_{o_{max}} < T_s \text{ (stagnation)}$$
$$< [T_a + \tau_{su} \, \alpha_{su} \, G_T/U_L]$$

$$< [10 + 0.8 \, (980)/8.0]$$
$$T_{o_{max}} < 108^oC$$

EXAMPLE B :   Black Plastic Swimming Pool Collector

Water flows under the entire absorbing surface of a 30-square metre black polypropylene swimming-pool collector with the same characteristics as the one in Figure 4.6.   The collector has a wall thickness of 2 mm and is unglazed.   The flow rate is such that U_f is 1500 W/(m^2 ^{o}C), $m_c \, c_c$ is 1000 W/^{o}C, T_i is 22°C and T_o is 27°C. The collector has an average heat loss coefficient (U_L) of 15 W/(m^2 ^{o}C) and the insolation (G_T) is 1030 W/m^2.   Find

  a) The useful energy collected and the collector
      efficiency if the ambient temperature is 20°C.
  b) Useful energy collected and the collector
      efficiency if the ambient temperature is 30°C.
  c) The "worst case" stagnation temperature, i.e.
      for G_T = 1030 W/m^2, T_a = 40°C, U_L = 15 W/m^2 ^{o}C.

Solution

  For this collector, $A_c = A_f$ and $\eta_F = 1.0$

a) $q_u = F_R [\tau_{su} \, \alpha_{su} \, G_T - U_L (T_i - T_a)]A_c$

Substitution into equation (5.8) gives the value of F_R for this example as

    F_R = 0.809
thus,
    q_u = 0.809 [1.0(0.93)1030 - 15(22 - 20)]30
    q_u = 23.25 kW

The collector efficiency is

$$\eta = q_u / (A_c \, G_T)$$

$$= 0.7524$$

b) $q = 0.809 \, [1 \times 0.93 \times 1030 - 15 \, (22-30)] \, 30$

$\qquad = 26.16 \text{ kW}$

and the collector efficiency is

$$\eta = 0.8467 \text{ (this can be } > 100\% \text{ when } T_a > T_i)$$

c) From equation (3),

$$T_{pmax} = T_a + \tau_{su} \, \alpha_{su} \, G_T / U_L$$

$$= 40 + 1030(1.0)0.93/15$$

$$= 104^O C$$

Since polypropylene will degrade and creep significantly at $104^O C$, it is essential that water be circulated through these collectors during hot days even if the heat is not needed for the pool.

APPENDIX 2

ANALYSIS OF TUBE-IN-PLATE COLLECTOR

INTRODUCTION

Probably the most common configuration for a solar collector
consists of a liquid flow channel attached to an absorber plate,
such as that shown in Figure 5.1.   The material used for the
absorber plate, the plate thickness and the tube spacing all have a
significant effect on the final cost of the collector.   The purpose
of this Appendix is to determine the relationship between the
collector efficiency and each of these factors.

MATHEMATICAL MODEL

The useful energy absorbed by a solar collector is given by

$$q_u = G_T \tau_{su} \alpha_{su} A_c - U_L A_c (T_p - T_a) \qquad (A1)$$

and the collector efficiency (η) is defined as

$$\eta = q_u/G_T \qquad (A2)$$

so

$$\eta = \tau_{su} \alpha_{su} - U_L(T_p - T_a)/G_T \qquad (A3)$$

In this equation only the average plate temperature (T_p) is
significantly affected by the choice of plate material, plate
thickness and tube spacing.   The temperature distribution along the
plate can be estimated by treating the plate as an extended surface
(or fin) such as that shown in the sketch on the next page.

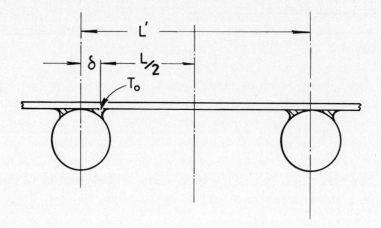

For this analysis it will be assumed that the tube spacing (L') is large relative to the tube diameter so that

$$L' = L$$

In that case, the one-dimensional, steady-state conduction equation becomes[*]

$$d^2T/dx^2 = - q_{GAIN}/k \qquad\qquad\qquad (A4)$$

where q_{GAIN} is the net heat gained (per unit volume) by the fin, or

$$q_{GAIN} = (G_T/t) - q_{LOSS} \text{ (per unit volume)}$$

If the insolation (G_T) and the losses (q_{LOSS}) are assumed to be uniformly distributed over the entire surface of the plate, then q_{GAIN} is a constant, and, from equation (A4),

$$d^2T/dx^2 = \text{constant} \qquad\qquad\qquad (A5)$$

The boundary conditions for this system are,

[*] In the regions of the plate close to the tube,
   two dimensional conduction effects become significant
   and this analysis is no longer valid.

(a) $x = 0$, $T = T_0$ T_{LIQUID}
(b) $x = L$, $dT/dx = 0$

With these boundary conditions, integration of equation (A5) gives the temperature distribution along the plate as

$$T_p(x) = T_0 + (q_s - q_L) [Lx - x^2/2]/(kt) \qquad (A6)$$

where $(q_s - q_L)$ is the net energy absorbed per unit area of plate. This energy will vary with the local plate temperature, but the variation will have only a small (second-order) effect on the collector efficiency, so $(q_s - q_L)$ will be assumed to be a constant in this analysis.

Further integration of equation (A6) gives the average plate temperature (T_p), i.e.,

$$T_p = T_0 + (q_s - q_L) L^2/(3kt) \qquad (A7)$$

Consequently, the collector efficiency in equation (A3) becomes,

$$\eta = \tau_{su} \alpha_{su} + U_L(T_0 - T_a)/G_T - U_L (q_s - q_L) L^2/(3kt) \qquad (A8)$$

or, for the same plate material and thickness the relationship between the efficiency and the tube spacing is

$$\eta = C_1 - C_2 L^2 \qquad (A9)$$

where C_1 and C_2 are constants.

Similarly, the only effect either the thermal conductivity or the thickness of plate has on collector efficiency is through the last term in equation (8). Thus, the effect of these variations on plate efficiency is given by

$$\eta = C_3 - C_4/t \qquad (A10)$$

and

$$\eta = C_5 - C_6/k \qquad (A11)$$

COMPARISON OF THEORY WITH EXPERIMENT

Speyer [13] measured the collector efficiencies of copper, steel and

aluminium collectors for a range of plate thicknesses and tube
spacings.   These data can be used to test the accuracy of equations
(A9), (A10) and (A12).   Equation (A12) indicates that when the
plate material and thickness remain constant, a plot of collector
efficiency against the square of the tube spacing (L^2) will be s
straight line.   Figure 4.7 shows that Speyer's data do form a
straight line for the copper and aluminium plates.   For the steel
absorber plates, the data did deviate significantly from the
straight line for values of L less than 50 mm, particularly for the
thinner plates.*   Similarly, Figures 4.8 and 4.9 show that Speyer's
data are linear functions of $1/t$ and $1/k$ and Figure 4.10 shows that
the data for all of the collectors tends to fall close to a straight
line when the combined parameter (L^2/kt) is used.

*This was probably because the thermal conductivity
 of the bond between the tube and the plate did not
 have a significantly lower thermal conductivity than
 the steel, so two-dimensional conduction effects were
 significant and equation A5 was no longer valid.

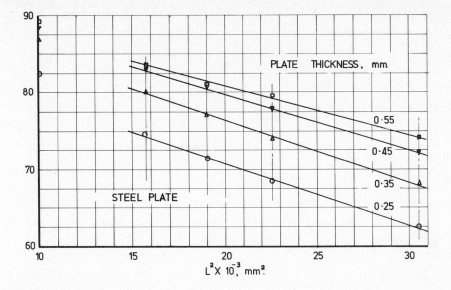

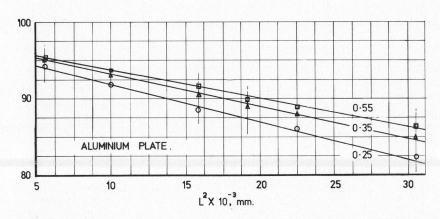

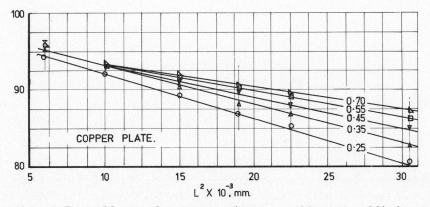

Fig. 4.7.   Effect of tube spacing on collector efficiency.

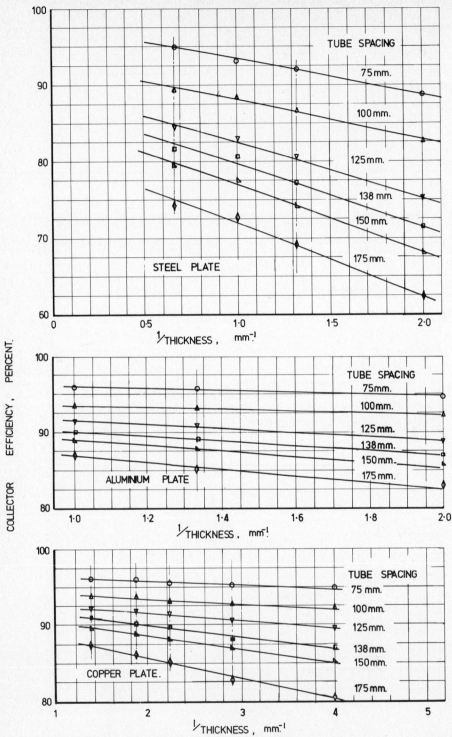

Fig. 4.8. Effect of absorber plate thickness on collector efficiency.

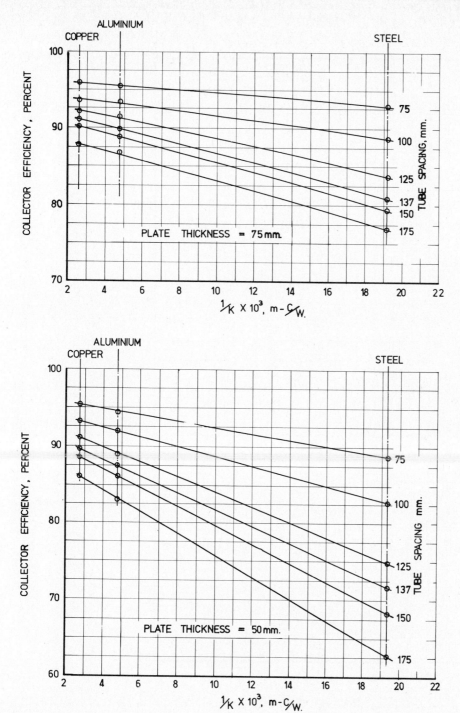

Fig. 4.9.   Effect of plate conductivity on collector efficiency.

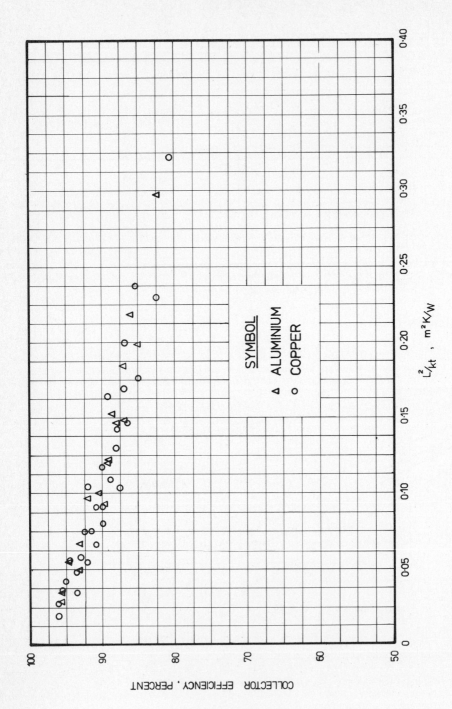

Fig. 4.10.   Some parameters that influence collector efficiency.

NOMENCLATURE

Symbol	Meaning
A_c	the total collector area (m^2).
A_f	the fin area (m^2)
c_c	the specific heat of collector fluid (kJ/kg $^{\circ}$C)
F_R	the "heat recovery factor" (dimensionless)
G_T	the instantaneous solar radiation incident on the collector (W/m^2)
k	the thermal conductivity (W/m $^{\circ}$C)
L	the effective tube spacing in a solar collector (m^2)
L'	the center-to-center distance between the tubes (m)
m_c	flow rate of collector fluid.
q_{GAIN}	the net heat gain (per unit volume) of the absorber plate (W/m^3)
q_u	the useful energy delivered by the collectors (W).
t	the thickness of the absorber plate (m)
T_a	the ambient temperature of the atmosphere ($^{\circ}$C)
T_i	collector inlet water temperature, $^{\circ}$C
T_o	collector outlet water temperature, $^{\circ}$C
T_p	the average temperature of the absorber plate ($^{\circ}$C).
T_s	the stagnation temperature for the solar collector ($^{\circ}$C)
U_f	the overall heat-transfer coefficient between the coolant and the absorber plate (W/m^2 $^{\circ}$C)
U_L	the overall heat-transfer coefficient between the absorber and the surroundings (W/m^2 $^{\circ}$C)
α_{su}	the solar absorptivity of the absorber (dimensionless)
η	instantaneous collector efficiency (Figure 4.5)
η_F	the fin efficiency
τ_{su}	the solar transmissivity of the cover (dimensionless).

REFERENCES

1.   Ratzel, A.C. and Bannerot, B. "Optimal Material Selection for
     Flat-Plate Solar Energy Collectors Utilising Commercially
     Available Materials", ASME-AIChE National Heat Transfer
     Conference, St. Louis, Mo. (August 1976).

2.   Grimmer, D.P. and Moore, S.W. "Practical Aspects of Solar
     Heating: A Review of Materials Use in Solar Energy
     Application", Los Alamos Report No. LA-VR-75-1752, Los Alamos
     Scientific Lab., Univ. of Calif. (1975).

3.   Colorado State University, "Solar Heating and Cooling of
     Residential Buildings: Design of Systems", U.S. Government
     Printing Office, Washington, D.C. (1977).

4.   "Qualification Test and Analysis Report to Marshall Space
     Flight Center (NASA) for SUNPAK^{TM} Model SEC601 Air
     Cooled Collector" Solar Energy Products Group, Owens-Illinois
     (August 1978).

5.   Chopra, P.S., "Reliability and Materials Performance of Solar
     Heating and Cooling Systems", Argonne National Lab. Report No.
     SOLAR/0906-79/70 (July 1979).

6.   Mavec, J.A., et al., "Reliability and Maintainability
     Evaluation of Solar Collector and Manifold Interconnections",
     Argonne National Lab., Report No. SOLAR/0902-79/70
     (March 1979).

7.   Waite, E., et al., "Reliability and Maintainability Evaluation
     of solar Control Systems", Argonne National Lab., Report No.
     SOLAR/0903-79/70 (March 1979).

8.   Cheng, C.F. "Preliminary Materials Assessment in Solar
     Demonstration Systems", Energy Technology Conference, Houston,
     Texas (November 1978).

9.   Moore, S.W., "Status Report on Solar Absorber Paint Coatings",
     Los Alamos Scientific Labs., Report No. LA-8897-SR (1981).

10. Boehm, R.F. and Swanson, S.R., "Solar Systems Design Handbook
     for Utah", Published by Utah Engineering Experimental Station,
     Univ. of Utah (June 1978).

11. Tabor, H. "Solar Energy Conversion", edited by A.E. Dixon and
     J.D. Lesley, Pergamon Press (1979), pp.253-286.

12. Private visit to Los Alamos by E. Baker (September 1982).

13. Speyer, E., "Solar Buildings in Temperate and Tropical
     Climates", Paper No.S8, New Sources of Energy Conference,
     Rome (1961). [Data also published by Szokolay, S.V., "Solar
     Energy and Building", Second Edition, London: The
     Architectural Press (1977), p.19].

CHAPTER 5

Collector Requirements
for some Specific Applications

INTRODUCTION

For a particular application, the selection of the appropriate
collector is dependent upon various parameters including the
physical environment, the working fluid, the desired temperature
rise and the flow rate.   In Chapter 2 heat balance calculations
were performed for domestic heating requirements.   In this chapter
the heat balance approach will be extended to some other
applications which use flat-plate or evacuated-tube collectors.
There is no one optimum collector design for all applications, and
different types of collectors are required for swimming pool
heating, for domestic water heating, for process heating, for space
heating and for comfort cooling.   The type of collector best suited
to each function will be described in this chapter.

SWIMMING POOL HEATING

Since the solar heating systems used for swimming-pools are probably
the simplest of all solar heating systems, they will be considered
first.   Small temperature rises with high volume-flow rates are
required for heating swimming pools.   Standard glazed and insulated
collectors are unnecessary for this application. Accordingly costs
may be reduced by dispensing with expensive containers for the
collector.   Furthermore, for the low temperature rises involved
selective surfaces give little benefit, and the use of less
expensive paints and plastics is perfectly appropriate.   An
additional penalty of glass covers is that a transmissivity loss of
at least 10% is common - a figure which increases strongly for
non-optimal inclinations.

A heat-balance analysis for a swimming pool in Sydney has been given
in Appendix 1.   The analysis is only for the month of November, but
the calculation may readily be repeated for other months.   The

results are very sensitive to windspeed at the pool surface and also
to the use of a pool cover.

Windspeed data available for a city such as Sydney usually pertain
to a height of 18 metres above the surface (Table 5.1).   To obtain
values near ground level in open country the one-seventh power-law
approximation for turbulent boundary layers is used.   For a height
of 0.3 metres above the ground, the one-seventh power law
approximation gives

$$v_{0.3} = 0.56 \ v_{18}$$

Pool fences, walls, bushes, or other obstructions which shelter the
pool from the wind will have a large effect on the wind speed.
Consequently, the velocity $v_{0.3}$ must be further divided by a factor
to allow for the degree of shelter.   For very windy environments
this factor approaches unity and for a completely sheltered
environment with negligible wind velocities the factor would tend to
infinity.   For a normal pool environment a factor of 4 is often
used, so a factor of 4 has been used in the example of Appendix 1.

The evaporation losses are generally the greatest of all the heat
losses from the pool.   If the pool is always covered no evaporation
will take place, but the effectiveness of the pool cover will be
reduced significantly if its upper surface is wet.   The radiation
and convection losses will not be reduced by a cover since the
temperature of the cover will be near water temperature.

Floating pool covers are often made of two layers of PVC separated
by strips of polyurethane foam, or of a poly-bubble arrangement.
Overnight the temperature of a covered pool will usually drop by
about $1^{\circ}C$ compared with a $2^{\circ}C$ drop for an uncovered pool.   This
represents over a 100 kWh heat loss per day in a small pool.

The simplest and possibly most cost-effective type of pool collector
uses a conventional corrugated metal-deck roof.   Water which
trickles from a perforated pipe laid along the ridge is heated as it
flows over the metal roof.   If the channels are left uncovered the
evaporation losses will be substantial.   The water will also

TABLE 5.1   Monthly Means of Meteorological Date for Sydney. [16]

Month	Max. Temp. °C	Min. Temp. °C	Average Temp. T_a °C	Wind Vel. v_{18} km/h	Vapor Pressure p_a mbar	Dew pt. Temp., T_d °C
Jan.	25.4	18.5	21.9	12.3	18.8	17.8
Feb.	25.4	18.6	22.0	11.6	19.2	18.2
Mar.	24.7	17.6	21.1	10.5	18.3	18.1
Apr.	22.3	14.4	18.4	10.2	14.9	15.4
May	19.8	11.4	15.6	10.5	11.9	12.9
June	17.1	9.1	13.1	11.6	10.1	10.8
July	16.7	7.9	12.3	11.5	9.6	10.0
Aug	18.0	8.8	13.4	12.1	9.5	10.1
Sept.	20.0	10.7	15.4	11.6	11.3	11.7
Oct.	21.8	13.3	17.6	12.3	13.1	13.3
Nov.	23.4	15.4	19.4	12.4	14.9	15.0
Dec.	24.7	17.2	21.0	12.3	17.6	17.1

accumulate dust and debris. An alternative is to use an inexpensive transparent cover on the roof.

A second inexpensive alternative, successfully exploited in the U.K. [1] uses the pool surroundings as collector surface. Plastic pipe may be laid in darkened concrete around the pool and in one application an adjacent tennis court was used as a collector.

The most successful solar pool heater is the fully enclosed extruded black plastic collector. The addition of carbon black to polyolefin resins gives excellent ultraviolet stabilisation. A number of well-designed proprietary pool-heating systems are available in Australia that use these black plastic collectors and the available evidence suggests that the life expectancy of these units is reasonable.

DOMESTIC WATER HEATING

A recent study by the CSIRO [2] found that solar energy is likely to capture over 60% of domestic water heating by the year 2020. Domestic water heating applications are characterised by a higher temperature rise requirement than swimming pool heating but with the delivery temperature not greater than 60°C. There is a significant danger from systems using efficient collectors that may approach temperatures of 95°C in pressurised water systems - causing scalding and steam release problems when the tap is opened. In addition, corrosion of domestic water tanks and systems can be quite rapid at these high temperatures.

The design specifications for the collectors used in domestic water heaters are quite different from those of swimming pool heaters. On the one hand, the type of unglazed, open system used for swimming pool heating has an efficiency characteristic which falls off too rapidly at higher temperatures (Figure 4.5) so it is unlikely that an unglazed collector would provide water at temperatures approaching 60°C over a sustained period. On the other hand, if the efficiency characteristic is too flat, the collector will continue to provide heat at high delivery temperatures approaching

$95^{\circ}C$ which could have disastrous implications for the system.   For this reason it is generally agreed that for Australian conditions the use of double-glazed collector panels is not appropriate for domestic water heating.   However, even though many designers believe that selective surfaces are also undesirable for domestic water heating, both selective surfaces and paint are used on commercial solar domestic hot water systems in Australia.

The collector area needed is determined by dividing the daily heating requirements by useful energy absorbed by the collecter each day

$$q = m_c c_w \, \Delta T$$

where ΔT is the difference between the temperature of the water leaving the collector (say 57°) and the mean mains supply temperature (Table 5.2).   For Sydney, the minimum collector area needed to heat 180 litres of water per day from from the monthly mean mains temperature to $57^{\circ}C$ is 2.60 m^2, the maximum 7.33 m^2 and the year-round average 4.09 m^2.   It is obviously feasible to provide more collector area and reduce the boost requirement from auxiliary energy sources, but the gain in useful energy does not generally keep pace with increased first cost.   The life-cycle cost analysis described in Chapter 11 can also be used to optimise solar domestic hot water systems.

For protection against frost damage, many overseas manufacturers prefer that the domestic water not pass through the collector itself.   In this case a hermetically sealed, secondary heat-transfer loop is incorporated between the collector and the hot storage tank, and either an oil or glycol fluid is used in this loop. These units are called closed-loop or closed-cycle systems. Corrosion problems are not as severe for such systems and all-aluminium construction is sometimes feasible.   Because these systems are more expensive and less efficient than the open-cycle systems, most Australian suppliers have favoured open-cycle systems. With open-cycle systems, the collector corrosion problems are potentially more severe and pressure requirements more rigorous. For these reasons the coolant passages in most open cycle units are

TABLE 5.2   Minimum Area Required for Each Month of the Year to Heat 180 Litres of Water from Mains Temperature to 57°C. (Sydney Data)

Month	Mean Mains Water Temperature °C	Average Heat required to heat 180 Litres (kJ/day)	Heat absorbed by collector [3] kJ/(m² day) (inlet temp. 35°C)	Minimum area of absorber required (m²)
Jan	22	26,269	9130	2.88
Feb	21.9	26,344	8570	3.07
Mar	20.9	27,094	8100	3.35
Apr	18.3	29,095	7590	3.83
May	15.1	31,550	6500	4.85
June	12.8	33,340	4550	7.33
July	11.8	34,090	6210	5.49
Aug	13.1	33,113	6560	5.05
Sept	15.1	31,550	8110	3.89
Oct	17.6	29,695	7900	3.76
Nov	19.4	28,233	9390	3.01
Dec	21.1	26,951	10380	2.60
Average		29,777	7,749	4.09

usually made from copper.

Although protection against electrolytic action between the tube and the plate is important, this problem has not prevented the use of aluminium plates into which copper tubes are pressed. In such designs, a thin film of thermal grease is sometimes used to improve the heat transfer and to provide some protection against electrolytic action. However, many experts are not convinced that the thermal greases that are currently available are likely to survive 20 years in a solar collector without significant degradation in performance [4].

Since the most promising application of solar energy is for domestic water heating [2], several organizations (e.g. University of NSW and SERI in Western Australia) are developing facilities for testing the long-term performance of complete systems. In April 1983, the Standards Association of Australia published a Draft Standard for these tests [5]. In the performance testing of systems, total systems are operated using a typical daily demand schedule over extended periods of many months and the operating characteristic of the system are monitored continuously. Table 5.3 [6,7] shows long-term performance data obtained from several different types of domestic hot-water systems. In this table,

$$f_R = [q_x - q_s]/q_x$$

and

$$\eta = [q_x - q_s]/[A_c \, G_T]$$

where q_x is the total electrical energy that would be consumed by a conventional hot water system under the same test conditions and q_s is the total electrical energy consumed by the system being tested. These results also give an indication how similar systems would perform in different climates. The first six systems were tested in the Arizona desert and the last three systems in Sydney. In both locations the pumped system, systems 1 and 9, was the most efficient of the systems tested, but the open thermosyphon system, systems 2 and 8, performed nearly as well. Selective surfaces were used on all four of these "top rating" systems. Unfortunately, the

TABLE 5.3   Overall Performance of Different Domestic Hot-Water Systems [3,6]

System	Collector Area (m^2)	$F_R \tau_{su} \alpha_{su}$	$F_R U_L$ (W/m^2°C)	Storage Capacity (kg)	Test Duration (Days)	f_R	Solar Fraction (f)	η
1. Drain Down (pumped system)	3.54	0.62	5.73	250	229	0.51	0.43	0.26
2. Thermosyphon (open)	3.98	N.A.	N.A.	300	229	0.49	0.41	0.22
3. Linear Concentrator (parabolic reflector)	7.81	0.33	1.48	250	226	0.47	0.38	0.11
4. Serpentine	4.56	0.59	5.16	310	203	0.45	0.34	0.15
5. Thermosyphon (closed)	4.25	0.58	5.28	250	210	0.41	0.31	0.16
6. Heat Pumps	-	-	-	197	169	0.46	-	-
7. Thermosyphon (open)	3.74	0.67	7.05	300	365	0.55	0.48	0.19
8. Thermosyphon (open)	3.76	0.68	5.18	300	365	0.63	0.65	0.25
9. Drain-Down (pumped system)	3.76	0.65	4.46	370	245	0.67	0.62	0.28

test conditions used at the two locations were quite different (e.g. test duration, demand curve, etc.) so a direct comparison of the two sets of test results is not possible.   Systems 2 and 8 (the thermosyphon system) were supplied by the same manufacturer and are an Australian-made system that is currently used widely throughout Australia as well as the U.S.   With the exception of system number 6, solar hot water systems similar to all of the other systems described in the table are currently in use in Australia.   System 6 has a reverse-cycle refrigeration unit mounted on top of the hot-water tank.   This system has only recently been introduced in the U.S. and has an installed cost of less than $1000 (U.S.).   It could have a promising future in Australia, particularly if it is able to use off-peak electricity.   The results of these tests were all slightly lower than, but within 8% of, the predictions of the f-chart analysis that will be described in Chapter 6.   Since the storage tanks were all located outside, the resulting increase in heat loss from these tanks probably could account for even this small difference.

PROCESS HEATING

The use of solar heat for industrial and agricultural purposes may become one of the most beneficial uses of solar energy for Australia.   The variety of such applications is endless.   Uses include crop drying, distillation, greenhouses, industrial air heating, industrial water heating, intensive fish farming, minerals processing, solar furnaces and steam generation.   The present discussion will be confined to water heating applications at temperatures below $100^{\circ}C$.

An American study [8] has shown that less than 5% of America's industrial process heating needs occur at temperatures below $100^{\circ}C$, and much of this low-temperature heat could be supplied more economically with heat-recovery systems added to existing non-solar systems.   However, in specific industries in Australia, the prospect for solar energy utilisation is much more promising.   The CSIRO study of Proctor and Morse [9] took the food processing industry as an example and found that 70% of the primary energy consumption was for processes operating at temperatures below 100°C. About 40% of the need was in the 60° to 80°C range. Typically at present, heat is generated in a central boiler and circulated to individual processes as high temperature water or low grade steam.

CSIRO have initiated a number of industrial projects aimed at solar energy utilisation in specific processes. One example of this is the Coca-Cola plant at Queanbeyan which uses a hot water spray over soft-drink cans freshly filled with ice-cold beverage. The warming of the cans prior to packing eliminates the risk of condensation which would be harmful to the cans and their packaging. The 77 m^2 flat plate collector used for this installation delivers water at temperatures up to 80°C. The collectors are an improvement on existing domestic designs to enable efficient operation at up to 80°C. The collectors used are double glazed and have a selective surface.

Some mining operations require hot water at temperatures between 60° and 100°C which low temperature solar collectors could supply. Studies in Western Australia [10] and the U.S. have considered the use of solar energy in the leaching processes used to extract uranium. Hot water at temperatures between 60° and 100°C is percolated through the ore and the reactive metals dissolve. The dissolved metals may then be separated chemically or by precipitation. The quantities of hot water required are very large indeed. In both countries, these mines are located in remote but sunny regions; the cost of conventional energy sources in these areas is high and the use of either flat-plate collectors or solar ponds would be economically advantageous.

A typical solar pond design would consist of a pool of water 1 to 2 metres deep having a blackened bottom. Plastics are well suited for this containment function. A salt is dissolved in the pond and under equilibrium conditions the density will increase with depth. The sun's rays pass through the water and are absorbed by the black plastic bottom. The increase in density with depth impedes convection from the bottom layers, and the poor thermal conductivity of the overlying brine reduces upward heat losses to the extent that temperatures of 70° to $90^\circ C$ have been measured in the lower layers of experimental ponds.   Heat is removed by circulating liquid from the bottom of the pond through a heat exchanger.   The principal advantages of the solar pond are its low cost per unit area and the fact that thermal storage is already built in.

SPACE HEATING AND COOLING

Since analyses of space heating and cooling systems are given in other chapters only a brief description of these systems will be given here.   Because of the collector area, the collector orientation and the storage volume requirements, solar space heating is most likely to be incorporated into new homes or buildings from the design stage.   Effectively, the collector and storage arrangements become part of the building.   Such an integrated approach precludes the marketing of separate components for retrofit purposes.   The two main collector types for space heating are those using air as a working fluid and those using a liquid.

The operating principles of water or liquid-heating collectors are identical with those for domestic water heating.   The main difference is one of size and complexity.   It is also likely that use of thermosyphon systems would not be feasible for most space heating applications.

Use of air as a working fluid entails a bulkier system than for a liquid but one which should have fewer corrosion problems and one which is potentially more simple to fabricate.   Glazing arrangements correspond broadly to those for the equivalent liquid systems, but the absorber plates are suitably shaped to ensure that the air flow is turbulent and thus the the heat transfer maximised.

Absorption cooling and Rankine-cycle vapour-compression systems have been the principal method of cooling by solar energy to date. Absorption systems using either lithium-bromide and water or ammonia and water have both been used, but the lithium-bromide system is the only type marketed at present.   An advantage of this working fluid is its low toxicity.   Temperatures required are in the range 75° to

$100^{o}C$ with optimum performance (giving a C.O.P. of 0.6) at $88^{o}C$. Clearly this temperature range is one which is possible for flat plate collectors. However, it is a demanding one. The flat plate collectors used by one of these systems are dimple-welded from stainless steel and have a good selective surface (see Table 4.4), but both copper and aluminium collectors are also available which have selective surfaces. Single glazing has generally been used although there is a significant incentive to use double-glazing with low-iron glass. The availability of double layer sealed panels provides convenient implementation of double glazing in existing collector boxes.

Although absorption cooling is possible with existing technology, solar driven Rankine-cycle vapour-compression refrigeration systems are under development in several demonstration projects in the U.S. In those systems solar heat is used to vapourise an organic fluid such as R-11. An analysis of working fluids and a useful reference list have been given by Wali [11]. The vapour is used in a Rankine cycle arrangement to operate a high-speed turbine which drives the compressor of a separate vapour compression. Such a unit has been run on the sub-$100^{o}C$ temperatures provided by flat plate collectors. These collectors usually have selective surfaces, good insulation, and are often double-glazed. As the temperature approaches $100^{o}C$ the use of either evacuated-tube or concentrating collectors must be seriously considered. Although these systems are more complex and more expensive, the increased temperatures which these collectors can attain would yield the higher Rankine cycle efficiencies which would probably be required for comfort cooling from solar energy to become really cost competitive. A system installed at Los Alamos Laboratories [12] uses flat plate collectors that are an integral part of the roof and one installed at the University of New Mexico [13] uses evacuated tube collectors. Both these systems have performed well for several years; however, their installation and maintenance costs have been quite high.

COLLECTOR TESTING

THERMAL PERFORMANCE TESTING

Standard test procedures for liquid-type flat-plate collectors have
been developed for all of the above applications [15].   The two
basic approaches in use are the instantaneous efficiency procedure
and the calorimetric procedure.

Instantaneous Efficiency

In the instantaneous efficiency method the mass flow rate of the
coolant (m_c), the insolation (G_T) and the collector inlet and outlet
temperatures (T_O and T_i) are all measured simultaneously and under
steady-state conditions.   The instantaneous efficiency ( ) is then
calculated from

$$ = m_c c_c (T_O - T_i)/G_T$$

The equipment needed for outdoor testing consists of a stand to hold
the collector, a large supply tank for the water, a smaller header
tank, a circulating pump and monitoring instrumentation.   The
header tank ensures a steady pressure feed for the inlet water so
that the mass flow rate remains sensibly constant over the test
period.   The supply and header tanks are well insulated and the
water is mixed thoroughly to insure that the temperature of the
water entering the collectors remains constant.

Usually, a tracking collector with a sighter is used to ensure that
the collector is perpendicular to the sun's rays.   The time
interval between readings must be long enough to ensure that
equilibrium conditions are reached.   The mass flow can be
determined by running the discharge water into a measuring tank.

Test readings are repeated for different inlet temperatures and the
characteristic collector curve of instantaneous efficiency as a
function of $(T_i - T_a)/G_T$ is plotted.

All readings should be taken under clear skys, within about two hours of solar noon and for nearly constant insolation. Testing should only be performed if the wind speed is less than 3 m/sec. An exception to this is when the collector has no cover; in this case tests should be systematically repeated under conditions of different average wind velocity.

Calorimetric Procedure

The calorimetric procedure is favoured by some equipment manufacturers as an alternative to the instantaneous efficiency test. This test measures the system performance over a full day. A closed system is used in which the time rate of temperature change $[(T_i - T_f)/\text{elapsed time}]$ and related to the insolation by

$$= m_T c_c \, (T_i - T_f)/(G_T \text{ x time})$$

The instantaneous efficiency procedure is often the most convenient approach and is described in Australian Standard 2535 [15].

EXPOSURE TESTING OF COLLECTORS

The long-term performance and reliability of solar collectors are important factors in their acceptance by the public and by industry. In addition to data obtained from new collectors under controlled laboratory conditions, the designer must also know how the system will perform after 10, 15 or 20 years and if this performance will degrade significantly after short-term exposure to stagnation conditions. Exposure testing is one of the tools used for this purpose [14]. Many American states (e.g. Florida and California) will not allow collectors that fail these tests to be marketed in those states. The test procedure accepted by most states in the U.S. for exposure testing of flat-plate solar collectors is described in Appendix 2. Although exposure testing cannot identify all of the likely long-term problems, it has proven to be effective in identifying many design faults and potential problems in solar collectors.

Solar collectors made by experienced collector designers are usually able to withstand the thermal stresses and thermal shocks produced during this exposure test.   The usual cause of collector failure detected during these tests is from outgassing of the various materials in the collector or from changes in material properties due to thermal or photodegradation.   Although many other examples could be cited [14], the conclusion is clearly that exposure testing can reveal some of the faults in a new design.   Consequently, exposure testing should be a part of any collector design and selection program.

APPENDIX 1

SWIMMING POOL ANALYSIS

Heat Balance for Open Air Swimming Pool in Sydney

In this analysis it will be assumed that no pool cover is used and it is desired to attain a pool temperature (T_p) of $25^{O}C$. As indicated in reference [16, 17] this is a good target temperature based on a wide survey of preferences. Pool heat gains and losses should be calculated for each month of the year. The objective is to determine the proportion (S) of pool surface area which must be replicated in flat-plate collector area to achieve the desired temperature of $25^{O}C$ during each month. As a corollary it will be possible to estimate the length of swimming season which a given collector area will allow.

As an example, a detailed analysis will be described for the month of November. Relevant meteorological data (with references) are as follows:-

    Average air temperature (T_a) [18] (Table 5.1) $= 19.4^{O}C.$
    Average dew-point temperature (T_d) (Table 5.1) $= 15.0^{O}C.$
    Average vapour pressure (p_a) (Table 5.1) $= 1.49$ kPa

    Saturated vapour pressure of air at
    pool surface at $25^{O}C$ (p_w) [19] $= 3.12$ kPa

    Average windspeed at surface of reasonably sheltered pool,
$$v = 0.25 \times v_{0.3}$$
$$v = 0.25 \times 0.56 \times v_{18}$$
$$v = 0.5 \text{ m/s}$$

Mean daily irradiation for horizontal surface (G_T, from Table 3.4, also reference [3])

$$G_T = 6.25 \text{ kWh/m}^2 \text{ per day}$$

Mean daily irradiation for inclination 34° to horizontal (Table 3.4 and reference [3])

$$G_T = 5.72 \text{ kWh/m}^2 \text{ per day.}$$

The heat gains and losses are as follows:

Convection Loss

The relationship for convective heat-transfer coefficient (h) that is used to calculate the convection losses from swimming pools is composed of a constant term to represent the natural convection and a term proportional to wind speed which represents the forced convection. That is,

$$h = 0.136 + 0.091 \text{ v} \qquad \text{kWh/(m}^2 \, ^{\circ}\text{C) per day}$$

Thus, since

$$q_c = h \, (T_w - T_a)$$
$$q_c = (0.136 + 0.091 \times 0.5) \, (25 - 19.4)$$
$$= 1.019 \quad \text{kWh/m}^2 \text{ per day.}$$

Evaporation Loss

Evaporation losses occur when the pressure of the water vapour in the air is lower than the vapour pressure of the pool water. The equation for heat loss by evaporation is

$$q_E = 1.85 \, h \, (p_w - p_a) \text{ x k}$$

where k represents the proportion of the time that the pool is not covered. (If there is no pool cover, k is unity). For example,

$$q_E = 1.85 \times 0.182 \, (31.2 - 14.9) \times 1$$
$$= 5.488 \text{ kWh/m}^2 \text{ per day.}$$

Radiation Loss

Radiation losses to the sky and surroundings occur when the water temperature is higher than that of the surroundings

$$q_R = 1.1 \times 10^{-9} \left[T_w^4 - e_a T_a^4 \right]$$

In the above equation T_w and T_a are in $^\circ K$. e_a is the apparent emissivity of the surroundings and is given as a function of dew point (T_d) near ground level:

$$e_a = 0.80 + 0.0038 \, T_d$$
$$q_R = 1.1 \times 10^{-9} \left[(298)^4 - 0.857 \, (292.4)^4 \right]$$
$$= 1.788 \text{ kWh/m}^2 \text{ per day.}$$

Make-up Water Loss

Heat is also required for the make-up water that replaces the water lost by evaporation.   The latent heat of evaporation of water is 0.694 kWh/kg.   The mass transfer (m_e) for the example is therefore

$$m_e = 5.488/0.694$$
$$= 7.908 \text{ kg/m}^2 \text{ per day.}$$

The heat required for make-up water is given by

$$q_M = m_e \, c_c \, (T_w - T_g)$$

where c_c is the specific heat of water (4.18 kJ/kg $^\circ$C) and T_g is the mains water temperature (19.4 $^\circ$C from Table 5.2), so

$$q_M = 0.051 \text{ kWh/m}^2 \text{ per day}$$

q_M will be taken as 0.1 kWh/m^2 day to allow for backwash, splashing, etc.

Heat Gain from Direct Radiation

Absorption coefficients for swimming pools vary between 0.8 and 0.95

for conventional pools in good condition.   A value of 0.9 will be
assumed here.   The proportion of the time that the pool is not
artificially shaded, by buildings, etc., is k'.   For the example k'
will be assumed to be unity (pool not shaded).

$$
\begin{aligned}
q_{DIR} &= 0.9 \times G \times k' \\
&= 0.9 \times 6.25 \text{ kWh/m}^2 \text{ per day} \\
q_{DIR} &= 5.625 \text{ kWh/m}^2 \text{ per day.}
\end{aligned}
$$

Heat Gain from Collectors

The collector efficiency decreases linearly with the difference
between pool temperature and air temperature.   For unglazed
collectors it is also a strong function of windspeed.   For
sheltered collectors an assumed efficiency (η) of 0.75 is reasonable
(Figure 4.6).   However, if the collectors are exposed to strong
winds the value should be reduced towards 0.5.   Since only a
proportion (S) of pool surface area is replicated in collector
area, S becomes a multiplier in the collector heat gain equation.

$$
\begin{aligned}
q_{COLL} &= \eta \times G_T \times S \\
&= 0.75 \times 5.72 \times S \text{ kWh/m}^2 \text{ per day} \\
q_{COLL} &= 4.29 \text{ S kWh/m}^2 \text{ per day}
\end{aligned}
$$

The pool heat balance, averaged over a 24 hour period, can now be
written:

$$
q_{COLL} + q_{DIR} = q_C + q_E + q_R + q_M
$$

For this example this gives:

$$
4.29 \text{ S} + 5.625 = 1.019 + 5.488 + 1.788 + 0.100
$$

and

$$
S = 0.65
$$

Thus, in this example, the area of solar collector should be 65% of
pool surface area to maintain a pool temperature of 25°C in Sydney
during the month of November.   These calculations show that the

main heat loss from the pool was by evaporation and if a pool cover had been used for only 50% of the time, no solar heating would have been required for that month.   It is also clear from the above calculations that the required collector area is particularly sensitive to wind velocity.   If the pool location is very windy, a larger collector area will be required.   The effect of both wind velocity and the use of a pool cover can readily be quantified in the heat balance calculations.

The above calculations for the month of November should then be repeated for each month of the year and may be presented in suitable form such as that of Figure 5.1.   It is then a simple matter of inspection to ascertain the length of swimming season which a given proportion of collector area (S) would permit in an average year.

If the solar heating system is to be used in conjunction with a back-up heater, it is possible to proceed from the heat balance calculations to determine the economically optimum collector area. The total solar energy supplied over the swimming season is calculated by summing the monthly contributions.   The total pool losses for the year are also obtained by summing monthly contributions.   The f-chart procedure of Chapter 6 and the life-cycle cost analysis described in Chapter 11 can be used to determine the economic optimum collector area.

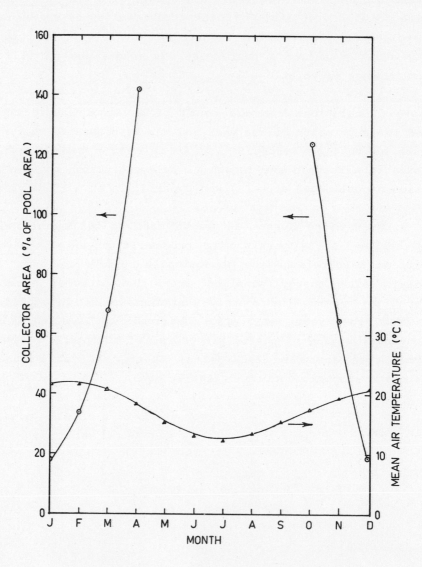

Fig. 5.1.   Collector area required as a percentage of pool area
to maintain a water temperature of 25°C (the pool is
not covered and it is located in Sydney).

APPENDIX 2
EXPOSURE TESTING

The exposure test performed at the Florida Solar Energy Center [20] is typical of the test procedure used throughout the U.S.   Although the primary purpose of the test sequence is to expose the collector to stagnation conditions over a 30-day period, several other short-term tests are necessary to help interpret the information obtained from the exposure test.   The test sequence includes the following:

1.   An incoming inspection to ensure that the collector is basically the same as the collectors sold to the public.

2.   A static pressure test at pressures about 10% above the manufacturers rated operating pressure to ensure that the collector can withstand these pressures without significant flow-path deterioration (swelling, stretching, etc.) or fluid leakage.

3.   Exposure to the sun and weather for a few days to precondition the collector so that subsequent performance tests will be more representative of its performance during actual operation.

4.   Measurement of the thermal time constant for the collector.   The thermal time constant of a collector is a measure of how long it takes a collector to respond to an external temperature change.

5.   A thermal performance test on the collector following ASHRAE Standard 93-77.   At least four tests are run at each of four different fluid inlet temperatures.   This test is similar to the one described in Australian Standard 2535 [15].

6.   An incident-angle modifier test is also conducted.

7.   A 30-day exposure test is conducted to verify the

consistency of the thermal performance tests and the integrity of collector under adverse conditions.

a.   This test usually takes longer than 30 days as each day must have a minumum cumulative radiation flux of 17000 kJ/m^2 in the plane of the collector aperture.

b.   These conditions must include at least one consecutive four-hour period with a minimum flux of 940 W/m^2.   (In water filled collectors this four-hour period must occur after the water has boiled out.

c.   For a five-minute period on three different days the collector is subjected to a water spray to produce the type of thermal shock that would occur during a rain shower.

8.   The thermal performance of the collector is again measured after completion of the exposure test to provide a quantitative measure of the extent of any degradation in performance.   It is this post exposure performance that is published and then used by designers to predict collector performance in actual applications.

9.   After completion of all other tests the collector is inspected for evidence of degradation or changes which could affect its functioning in normal service.   This visual inspection also helps identify the causes of performance changes.

NOMENCLATURE

Symbol	Meaning
A_c	the glass area of the solar collector (m^2)
c_c	the specific heat of collector fluid (kJ/kg $^\circ$C)
e_a	the apparent emissivity of the surface instantaneous collector efficiency
f	the solar fraction
f_R	the fractional energy savings of the system compared to a conventional electric hot water system
G_T	the instantaneous solar radiation incident on the collector (W/m^2)
h	the convective heat transfer coefficient on the exposed surface of the collector (W/m^2 $^\circ$C)
m_c	flow rate of collector fluid.
m_T	the thermal storage capacity of the system (kg of water/m^2 of collector area).
p	pressure (MPa)
p_a	the vapour pressure (MPa)
p_w	the saturated vapour pressure (MPa)
q_s	the total electrical energy consumed by the system
q_x	the thermal energy drawn from the system as hot water
S	the ratio of the collector area to the pool surface area
T_a	the ambient temperature of the atmosphere ($^\circ$C)
T_f	the temperature of the water in the system at the end of the test ($^\circ$C)
T_i	collector inlet water temperature ($^\circ$C)
T_i	the initial temperature of the water in the system ($^\circ$)
T_o	collector outlet water temperature ($^\circ$C)
v	the wind velocity (m/sec)

subscripts

0.3	refers to a height 0.3 metres above the surface
18	refers to a height 18 metres above the surface
c	the convective component
c	the coolant
COLL	refers to the collectors
d	at the dew point
DIR	the direct radiation
E	refers to evaporation
g	refers to mains water
M	refers to make-up water
R	the radiative component
w	refers to the pool water

REFERENCES

1.   Burra, J.P.M., "The Pool Surroundings and Terraces as Solar
      Collectors", U.K.- I.S.E.S. Practical Experiences with Solar
      Heated Swimming Pools (1978).

2.   Musgrove, A.R. del, et al., "Exploring Some Australian Energy
      Alternatives using MARKAL", Division of Energy Technology
      TR2, Lucas Heights Research Laboratories, N.S.W. (1983).

3.   Morrison, G.L., Sapsford, C.M. and Litvak, A., "Solar
      Insolation Data for Sydney", Univ. N.S.W. Report No.
      1979/FMT/2 (1979).

4.   Private communication with E.M. Fry, Reliability Department,
      Bell Telephone Laboratories, Whippany , NJ (1969) and Schertz,
      W.W., Solar Applications Manager, Argonne Labs, Argonne,
      Illinois (1983).

5.   "Household Solar Water Heaters, Method of Test for
      Performance", Standards Association of Australia (Draft
      Standard) (April 1983).

6.   Balon, R.J., Wood, B.D. and Nelson, D.J., "Performance
      Testing and Rating of Solar Domestic Hot Water Systems",
      CR-82030, College of Engineering and applied Science,
      Arizona State University (August 1982).

7.   Morrison, G.L. and Sapsford, C.M., "Long-Term Performance of
      Thermosyphon Solar Heaters", vol. 30, Solar Energy, (1983)
      pp. 341-50.

8.   Inter-Technology Corporation, Analysis of the Economic
      Potential of Solar Thermal Energy to Provide Industrial
      Process Heat, Vol.1, p.53 (Feb. 7, 1977).

9.   Proctor, D. and Morse, R.N. Solar Energy for the Australian
      Food Processing Industry. International Solar Energy
      Congress, U.C.L.A. (1975).

10.   Saunders, D.W. Industrial and Mineral Applications for
      Solar Energy. I.S.E.S. Symposium, Melbourne (Nov. 1976).

11.   Wali, E., "Optimum Working Fluids for Solar Powered Rankine
      Cycle Cooling of Buildings", vol. 25, Solar Energy, (1980)
      pp. 235-41.

12.   Solar Energy Group, "Solar Heating and Cooling System,
      National Security and Resources Study Center", Los Alamos
      Sci. Lab Report No. LASL 80-38 (Dec. 1980).

13.   Wildin, N.W., University of New Mexico, Albuquerque, N.M.
      (private communication during visit to Univ. of New Mexico by
      E. Baker in Sept. 1982).

14.   Baker, E., "Exposure Testing of Flat-Plate Solar Collectors", ISES Conf. (ANZ Sect.) "Solar Energy at Work" (Nov. 1981).

15.   "Glazed Flat-Plate Solar Collectors with Water as the Heat-Transfer-Fluid -- Method of Testing", Australian Standard 2535.

16.   Ministry of Housing and Local Government, "Swimming Pools", Design Bulletin No. 4, H.M.S.O., London (1962).

17.   Sheridan, N.R., "The heating of Swimming Pools; Solar Research Notes No. 4, University of Queensland (1972).

18.   Commonwealth Bureau of Meteorology, Sydney (1975).

19.   Keenan, J.H. and Keyes, F.G. Thermodynamic Properties of Steam, Wiley (1936).

20.   Florida Solar Energy Center document "Test Method and Minimum Standards for Solar Collectors", F.S.E.C. 77-5 (June 1977).

CHAPTER 6

Thermal Energy Storage

INTRODUCTION

The storage system allows heat collected during periods of excess sunshine to be retained until needed during the night, or during intermittently cloudy daytime periods. Poorly designed thermal storage can lead to inefficient system operation and to costly system failures. In fact, the plague of tank failures in the late 1940's, described in Chapter 1, virtually destroyed Florida's entire solar hot-water heating industry. Consequently, the design of the heat-storage system ranks close in importance to the design of collectors in solar heating systems. If the thermal storage is too small for the collector system, the storage temperature will increase, leading to greater heat losses and more rapid system degradation. On the other hand, if the storage system is too large it will be difficult to heat.

Any practical storage system must be inexpensive to buy, to install, to operate and to maintain. In addition, it must have few energy losses, and it must be able to economically store and deliver the heat where and when it is needed. In the following sections, sensible-heat storage systems will be described in detail, since, at present, these are the only commercially available systems that meet these requirements. Change-of-phase type storage systems will also be discussed since significant research is underway in an attempt to make these systems both reliable and economic.

SENSIBLE-HEAT-STORAGE SYSTEMS

With sensible-heat storage the quantity of heat stored is proportional to the temperature rise of the storage medium. The heat capacity of a material is a measure of its ability to store sensible heat. The heat capacities of several materials are listed in Table 6.1, in terms of both mass and volume.

TABLE 6.1   Heat Capacities for Common Materials

Material	Heat Capacity	
	kJ/(kg $^{\circ}$C)	kJ/(m^3 $^{\circ}$C)
Air	1.0	1.2
Water	4.2	4200
Wood	2.7	1600
Steel	0.49	3800
Concrete	0.88	1800
2-4 cm rock	0.80	1300
Brick	0.84	1350

Rock and water are used extensively since they are inexpensive and readily available.   They both have high heat capacities and good heat transfer characteristics; but, as the table indicates, water has three times the heat capacity of the same volume of rock.

Water-Storage Systems

Water has a heat capacity of 4200 kJ/m^3 $^{\circ}$C, so if a storage tank at 90°C were cooled to 35°C, 231,000 kJ of heat would be removed from each cubic metre of water in the tank.   This is of the same order as the energy required during an average winter day for heating a typical 140 m^2 residence in Sydney.   However, to reduce the heat losses, the optimum storage temperature range for an actual system would be somewhat less than 55°C and more than one day's storage would probably be required.   Consequently, the storage capacity that would be required for a home such as this would be closer to 4 m^3 (i.e. 4000 kg of water).

Water-storage tanks are usually made from steel, concrete or fibreglass reinforced plastics.   Since steel tanks tend to rust, they must be lined and they are usually protected with a sacrificial anode.   Fibreglass tanks are usually not suitable for these applications due to the high operating temperatures and the temperature cycling experienced by these tanks.   Cement tanks do not corrode and they are inexpensive but they must have a protective coating to reduce water permeation and, since they are loaded in tension, they have a tendency to fracture.

Water storage systems have several limitations on their operating temperatures. In regions where freezing is not a problem, water from the storage tank may be circulated directly through the solar collector and returned to the tank. Since pressurised systems are expensive, non-pressurised systems should be used where possible. Even with pressurised systems, the water temperature should be kept below 100°C. and the system should be vented to the atmosphere.

In a forced-convection home-heating system, if the temperature of the air supplied to the space to be heated drops below 30°C, the supply air will feel cool to the occupants. Also, the ratio of power required to drive the fans to the energy supplied to the space will become excessive. Consequently, with a forced-convection system the minimum useful temperature for the water in the storage tanks is around 35°C. Use of a high storage temperature will both increase the heat losses from the system and cause the average collector efficiency to decrease. With a natural convection system, such as those that use skirting-board heaters, the minimum useful storage temperature can be reduced slightly below 30°C, so these systems are able to deliver more of the stored heat to the space. With a skirting-board heating system the size of the water storage system can be reduced by about 10%; however, at these low storage temperatures, the rate at which the heat is transferred is quite low, so these skirting board air heaters are much larger for solar heating systems than for conventional systems. The effect is a tradeoff in which the heat exchanger costs increase while the operating costs decrease since fans are not required.

In an optimisation study designed to minimise the cost of (and to maximise the savings from) solar heating systems, Lof and Tybout [1] found that the capacity of the hot water storage system should be between 50 and 75 kg of water per square metre of collector area. Figure 6.1 shows the effect storage capacity has on solar heating systems. Increasing the storage capacity above 75 kg of water per square metre of collector area has very little effect on the fraction of the heating load that will be supplied by the solar heating system.

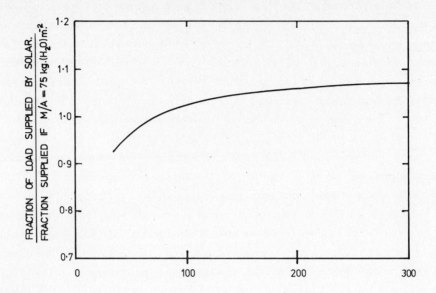

Fig. 6.1. Effect of storage capacity on the solar fraction [1].

Insulation

Some of the energy stored in the tank will be conducted through the walls of the tank and lost to the surroundings. If the tank is located inside the building, the heat lost may usefully warm that part of the building, but, since the rate of heat loss is not controlled, the area could become uncomfortably warm. In addition, during the summer, the heat from the tanks would increase the amount of cooling required. Consequently, even when located inside the structure, hot-water tanks should be well insulated. A thermal resistance (or R-value) of at least 2 m^2 $^{\circ}C/W$ for the insulation should be used for all types of thermal storage containers.

If buried tanks are used, insulation and strength may be the key problems. Moist soil is a good thermal conductor and moisture in the ground can penetrate many insulating materials making them useless. Also, soil and traffic loads create special structural problems for buried tanks.

Stratification

Since a fluid expands as its temperature increases, thermal stratification will normally occur in stored fluids. When stratification occurs, the hot fluid remains near the top of the storage container and the cooler fluid settles to the bottom. By using the warmer fluid at the top of the container for room heating, it is possible to recover a larger portion of the heat in the container. To take advantage of stratification, the fluid to be heated in the collector should be withdrawn slowly from near the bottom of the storage container (the cold end) and returned slowly at the top, otherwise fluid mixing will occur. Unfortunately, water storage systems are difficult to stratify and in a solar home-heating system, the difference in water temperature in a typical 4000 litre tank might be only $3^{O}C$. The amount of stratification in water storage systems can be improved by the use of baffles or multiple-tank systems which obstruct convection and fluid mixing. Since these modifications usually increase the cost of the system they may not be justified on economic grounds. Also, the resulting reduction in fluid velocity through the collector will cause the temperature difference between the inlet and the outlet of the collector to increase. The effect will be a reduction in the collector efficiency (i.e. a reduction in F_R) which will tend to nullify the advantages of stratified storage over well-mixed liquid storage systems.

In air systems, significant stratification is relatively easy to attain. Thus in air systems, stratification will reduce the advantage that water systems appear to enjoy over air systems due to the relatively high heat storage capacity of water [2].

EXAMPLE 1: Stratified Storage

For a particular application the operating conditions for a stratified-storage solar hot-water system without a heat exchanger are given as

$$m_c\ c_c = 45\ W/^{O}C \text{ per } m^2 \text{ of collector area,} \qquad G_T = 820\ W/m^2$$

storage size (m_T) = 75 kg (water)/m^2 of collector area

T_{so} (original storage temperature) = 25°C, T_a = 10°C

F_R = 0.853, $\tau_{su}\alpha_{su}$ = 0.86, U_L = 6.0 W/m^2 °C

Find

a.   The average temperature of the water in the storage tank after six.hours operation under these conditions.

b.   The total heat stored by the system per square metre of collector area.

Solution

a.   The turnover time (t_s) for the water in the tank is the time required for all of the water in the storage tank to make one pass through the collector system.   Thus, t_s is the ratio of the thermal capacity of the water in the tank divided by the heat transported to the tank from the collectors, or

t_s = 75(4190)/45
   = 6983 s (1.94 hours)

First turnover

From equation (6.1)

q = $F_R[\tau_{su}\ \alpha_{su}\ G_T - U_L(T_i - T_a)]\ t_s$
q = 0.853[0.86(820) - 6(25 - 10)]6983
q = 3664.4 kJ/m^2

and the temperature (T_s) of the water in the tank at the end of this turnover period is

T_s = T_{so} + $q/(m_T c_c)$

T_s = 25 + 3664.4/(75 x 4.19) = 36.66°C

Second turnover

$$q = 3246.6 \text{ kJ/m}^2, \quad T_s = 46.99 \text{ }^o\text{C}$$

Third turnover

$$q = 2877.5 \text{ kJ/m}^2, \quad T_s = 56.15^o\text{C}$$

Time remaining (t_{xs})

$$t_{xs} = 6 \text{ hours } - 3(1.94 \text{ hours/turnover})$$
$$= 0.18 \text{ hours } (651 \text{ seconds})$$

$$q_{xs} = 237.8 \text{ kJ/m}^2, \quad T_s \text{ (final)} = 56.91^o\text{C}$$

b)   The total heat collected by the system is

$$q(\text{total}) = 3664.4 + 3246.6 + 2877.5 + 237.8$$
$$= 10025 \text{ kJ/m}^2 \text{ of collector area}$$

EXAMPLE 2:   Well Mixed Storage

The hot water system in Example 1 is converted to a system with well-mixed storage.   For the insolation and ambient temperatures given below, find the final storage temperature and the total energy gained by the tank between 9 am and 3 pm.   The average storage temperature between 9 and 10 am (T_{i1}) is 25°.

Time	0900	1000	1100	1200	1300	1400
G_T, W/m^2	234.2	543.7	727.4	822.6	898.2	877.3
T_a, ^{o}C	10.4	12.9	14.5	17.6	18.2	18.4

Solution

For the first hour the average inlet temperature (T_{i1}) is 25°, so the useful energy collected during that hour (q_1) is

$$q_1 = F_R[G_T \cdot a_{su}\tau_{su} - U_L(T_{i1} - T_a)] \times 1 \text{ hour}$$

$$= 0.853[234.2(0.86) - 6.0(25 - 10.4)] \times 3.6$$
$$= 349.5 \text{ kJ/m}^2$$

and the average temperature in the water tank at the beginning of the second hour (T_{i2}) is

$$T_{i2} = T_{i1} + q_1/(m_T c_c)$$
$$= 25 + 349.5/(75 \times 4.19)$$
$$= 26.11^\circ C$$

For the second hour,

$$q_2 = 1193 \text{ kJ/m}^2, \quad T_{i3} = 29.90^\circ C$$

For the third hour,

$$q_3 = 1637 \text{ kJ/m}^2, \quad T_{i4} = 35.11^\circ C$$

For the forth hour;

$$q_4 = 1850 \text{ kJ/m}^2, \quad T_{i5} = 41.00^\circ C$$

For the fifth hour,

$$q_5 = 1952 \text{ kJ/m}^2, \quad T_{i6} = 47.21^\circ C$$

For the sixth hour,

$$q_6 = 1786 \text{ kJ/m}^2, \quad T_{i7} = 52.89^\circ C$$

Thus, the temperature in the storage tank after six hours operation is $52.89^\circ C$ and the total useful energy collected by the system was 8768 kJ/m^2. In this example the tank temperature was very low ($25^\circ C$) at the beginning of the day which gave a high collector efficiency during the first few hours of operation. Also, since no water was removed from the tank during this six-hour period, the temperature reached by the end of the period was the maximum possible for these conditions.

Rock-Bed Heat-Storage Systems

Unlike hot water, hot air cannot be economically stored for long
periods due to the low volumetric heat capacity of air (see Table
6.1).   Consequently, in most hot-air systems the heat is usually
transferred to well insulated rock-filled containers such as that
shown in Figure 6.2.   Plenums at the top and bottom of the
container allow the air to flow uniformly through the rock bed.

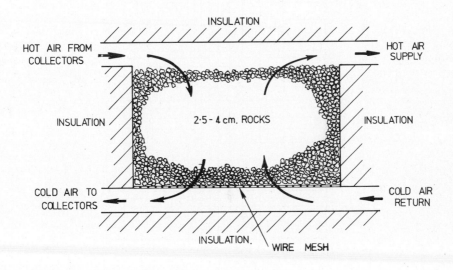

Fig. 6.2.   Rock-bed heat-storage unit.

The rocks used in storage units should be washed to remove dust that
may clog the filters and the rocks should also be fairly uniform in
size to prevent them from packing tightly.   Tightly packed rocks
will cause a non-uniform air flow across the unit which will reduce
the thermal efficiency of the storage unit and increase the fan
power required to force the air through the rock bed.   Round rocks
between 2.5 and 4 cm diameter are preferred [2], but sizes between 2
and 7 cm have been used successfully.   Rocks smaller than 2 cm
diameter tend to pack too tightly and the interior of rocks larger
than 7 cm in diameter may never be heated.   Although rounded rocks

are preferred, crushed gravel has been used successfully in regions where rounded rocks were too expensive. Approximate "rules-of-thumb" for the system parameters that are critical to the design of solar heating system are given in Table 6.2.

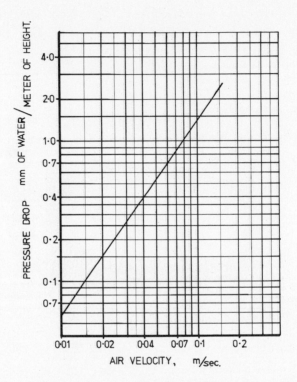

Fig. 6.3.   Measured pressure drop through a rock bin
             with 2 to 4 cm diameter rocks.

Another consideration with rock-storage units is the relative height of the container. Most successful beds are between 1.3 and 4 metres high. Tall, slender containers give better stratification than short containers and a minimum depth of 0.75 has been recommended [2]. The superficial air velocity for slender containers is greater than for containers with large cross sections. Consequently, as Figure 6.3 shows, the pressure drop and thus the power required for the fan will be much greater for slender containers. So to reduce this pressure drop, large diameter rocks should be used with slender containers. In several installations

TABLE 6.2   Rules of Thumb for Sizing Solar Heating System Components [2]

Collector tilt	Local Latitude + 15°
Collector flow rate	
Air	7.5 to 10 litres/s per m^2 of collector
Water	0.14 l/s per m^2 of collector area
Storage tank size for water	60 to 100 litres/m^2 of collector area
Pebble bed storage size	0.15 to 0.3 m^3 of rock per m^2 of collector
Rock depth (in direction of air flow)	1.2 to 2.4 m
Pebble size (smooth rocks preferred)	2.5 to 4 cm diameter
Duct insulation (min.)	25 mm fibreglass
Pressure drops:	
Pebble bed	2.5 to 7.5 mm of water
Air-cooled collector	
($\simeq$ 4 m long)	5.0 to 8.0 mm of water
($\simeq$ 6 m long)	8.0 to 13 mm of water
Water-cooled collectors	3.5 to 70 kPa per collector module
Ductwork	6.7 mm of water/100 m of duct length
Velocities	
Pebble bed (superficial velocity)	0.06 to 0.12 m/sec
Water pipes	1.5 to 2.5 m/s

in the U.S., the energy required to run the fan significantly reduced the overall efficiency of the hot-air solar heating system [3].

Since much of the total volume in the storage bin is not occupied by rock (approximately 40% is air space) and both stratification and usage factors will vary significantly between users, the exact calculation of the size of a rock storage unit is rather complex. Again, simulation studies indicate that the optimum volume of the storage bin will usually be between 0.15 and 0.3 m^3 (or between 200 and 500 kilograms of rock) per square metre of collector for solar heated dwellings.

Air Flow in Rock Beds

Since about 60% of the cross-sectional area of a rock-storage bed is occupied by rocks, the actual air velocity between the rocks will vary significantly with location within the bed. Consequently, an average (superficial) velocity is used to size the components of these systems. The "superficial" air velocity through a rockbed is defined as the air-flow rate (in m^3/sec) divided by the cross-sectional area of the storage bed (in m^2). It is this superficial air velocity that is used in Figure 6.3. In general, if the superficial air velocity is less than 0.05 metres per second, the heat transfer between the air and the rocks will be poor and the cross-sectional area of the bed will be unnecessarily large. However, if the superficial velocity is too high, the pressure drop through the rock bed may be very large and excessive power will be needed to drive the fans. Normally, the system should be designed to deliver a superficial air velocity of between 0.06 and 0.12 m/sec (see Table 6.2). Figure 6.3 shows that for these superficial air velocities the pressure drop through a 2-metre high rock bed would then be less than 4 millimetres of water.

As an example, suppose the collector area of a system is 50 square metres and the air-flow rate through the collector is 0.64 cubic metres per minute per square metre of collector. Then, air will enter the storage unit at the rate of 32 cubic metres per minute.

The smallest practical rock-storage system would use the maximum superficial air velocity of 8 metres per minute (0.133 m/s, Table 6.2) and the cross-sectional area of the tank would be 4 square metres.   Since 0.15 cubic metres of rock storage per square metre of collector are required for this "minimum" system (Table 6.2), then a 7.5 cubic metre rock bed will be needed.   For this case, the bed would be nearly 2 metres high and the pressure drop through the bed would be about 4 mm of water (Figure 6.3).   If 2 metres is too high for the existing space, a lower superficial air velocity might be used.   If a superficial velocity of 0.1 metres per second were selected, the rock bed would then be 2.3 metres on each side and 1.4 metres high.   In this case, the pressure drop through the bed would be about 2 mm of water.

Special Considerations for Rock Storage Units

Air leaks

Air leakage from the system is a source of heat loss. Unlike water leaks in the hot water system, air leaks are not always obvious and heat losses can become excessive before the owner realises that the system needs to be repaired.   Even when air leaks are known to exist, locating these leaks can be quite difficult.

Evaporative cooling

In regions where the air tends to become hot and dry (i.e. desert and similarly arid areas) evaporative cooling techniques can be used to cool homes.   Water sprayed over the rocks will evaporate and cool the air circulating through the rocks.   However, with evaporative cooling systems, condensation on the rocks can cause mildew and odours in the system.

HEAT STORED DURING PHASE CHANGE IN A MATERIAL

When a material changes phase (such as from a solid to a liquid) its temperature is nearly constant during the process and a considerable amount of heat is stored in the material.   For example, 335 kJ of heat are stored when 1 kg of ice melts; however, to store the same amount of energy in a sensible-heat-storage system would require nearly 3 kg of water and a recoverable temperature rise of over 25°C.   Consequently, phase-change materials offer the promise of smaller   storage   volumes   and   higher   collector   performance efficiencies.

Materials also absorb large quantities of energy when they change from the liquid to the gas phase; but, vapour storage requires either very large storage tanks, or storage at high pressure, or both.   Consequently, the liquid-vapour transition is not usually considered for home storage systems.

Since   liquid-to-solid   transition   occurs   at   a   characteristic temperature (water freezes at 0°C) most substances cannot be used for phase-change storage as their melting point is either higher or lower than needed for home heating.   The melting point for such materials should also not exceed 50°C since this would require high collector temperatures, resulting in excessive heat losses.   Also, if the melting point of the material is less than 3°C, high supply air velocities would be required that would create uncomfortable drafts in the space being heated.   In passive heating applications, this temperature range should be reduced to between 21° and 24°C. For space cooling the most desirable storage temperature range is between 5 and 18°C, so a different phase-change material may be required.

Various investigators have been studying phase-change materials for over 30 years [4] and have identified many possible compounds.   The most commonly mentioned compound, Glauber Salt, melts at 32°C, can be supplied in large quantities and at reasonable prices.   However, like most of these compounds, Glauber Salt tends to separate into various subcomponents after it has been frozen and thawed several times [5].   Once separation has occurred, the melting point of the material will change significantly.   For example, when first installed a salt may begin to melt at 35°C but, after it has been cycled a few hundred (or thousand) times the melting temperature may drop to 24°C.   In this case, when used for a forced air heating system, the air entering the living area would be about 19°C and would feel quite cool.

Computer simulation studies (e.g. [6]) indicate that between 7 and 25 kg/m^2 of collector area is the optimum mass for phase change materials in solar heating systems.   These studies also found that, because of the large surface area required to transfer the stored heat to the supply air, the storage volume saved by using phase-change materials is considerably smaller than expected.   In addition, the problems of packaging these salts to provide both the required surface area and the long life needed for these systems have not yet been solved [7].

A comparative analysis of sensible and latent-heat storage for home heating has been conducted by the Solar Energy Research Institute in the U.S.A. [8].   The purpose of the analysis was to identify the performance and economic advantages of using latent-heat storage systems.   No allowance was made for sub-cooling or degradation of the phase-change material so its conclusions tended to be biased in favour of latent-heat storage.   The major conclusions for home-heating use included the following:

1.   For air-based systems salt-hydrate latent-heat storage offers a four-to-one reduction in storage volume over rock-bed storage.

2.   For liquid-based systems, salt-hydrate latent-heat storage offers less than a two-to-one reduction in storage volume over hot water storage.

3.   Constant-temperature operation during phase change provides no operational advantage, and the volume reduction is the only advantage offered by latent-heat systems.

4.   The distinction between air- and liquid-based systems is far more important than that between sensible- and latent-heat systems.

Consequently, unless space is at a premium, the extra cost and the uncertain reliability associated with these latent-heat storage systems will probably exclude these units from the solar market for many years.

If a phase-change storage system is considered for any application the following points should be discussed:

1. Is the temperature at which phase-change occurs suitable for this application?

2. Can the system be used for both cooling and heating?

3. Is the phase-change material packaged effectively for heat transfer?

4. Does the performance of the system degrade significantly with time? The supplier should provide test data that ahows the degree of degradation of the system after 10,000 or more temperature cycles.

5. Since chemical changes are a function of both time and temperature, reliable life-test data will not be available for these systems for many years. Consequently, the warranty for the system should cover a significant portion of the total expense that will occur if the system must be removed and replaced with a different one.

In summary, the onus should be on the supplier to prove that his system is suitable for that application.

SIZING SOLAR HEATERS

THE DUFFIE-BECKMAN-KLEIN PROCEDURE

Development of the Theory

Duffie, Beckman and Klein [9, 10] developed a general design method for solar-heating systems that is based on an extensive sensitivity analysis of many different systems.   The results were presented in the form of a simple graph which uses monthly average meteorological data to size solar heating systems.   This method was used to develop the "rules of thumb" in Table 6.2 that are useful for sizing the various components of the system.   To demonstrate the use of this method, the home-heating system in Figure 6.4 will be sized to provide the optimal solar fraction (f) for a home in Sydney.   The solar fraction is defined as the fraction of the total heating load provided by the solar-heating system during the time period.

Duffie, Beckman and Klein developed this procedure for solar heating systems by varying each of the system parameters individually and observing the effect on the performance of the system and on the solar fraction (f).   More than 300 simulations were made, each of which estimated the month-by-month system performance using data for an average year in Madison, Wisconsin.   Their results were presented in graphical form as the f-chart in Figure 6.5.   This relationship between X, Y and f is also given by equation (6.1) [10].

$$f = 1.029\ Y - 0.295\ Y^2 + 0.0215\ Y^3 - 0.065\ X + 0.0018\ X^2 \quad (6.1)$$
$$\text{for } 0 < Y < 3 \text{ and } 0 < X < 18$$

For systems that use air for the heat-transfer fluid in the collector the relationship is

$$f = 1.040\ Y - 0.245\ Y^2 - 0.0095\ Y^3 - 0.065\ X_0 + 0.00187\ X_0^2 \quad (6.2)$$
$$\text{for } 0 < Y < 3 \text{ and } 0 < X_0 < 18$$
where
$$X_0 = X\ (mc/0.409)^{0.3}\ (V/10.1)^{0.28}$$

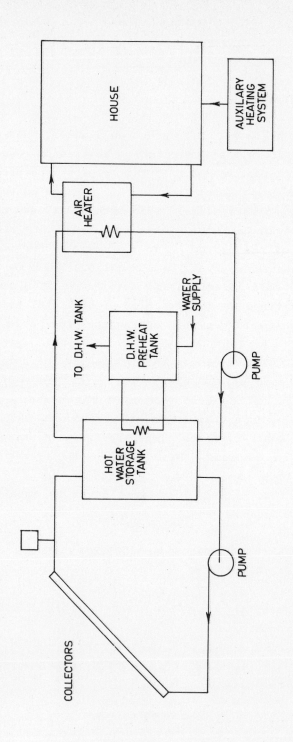

Fig. 6.4.   Home heating system.

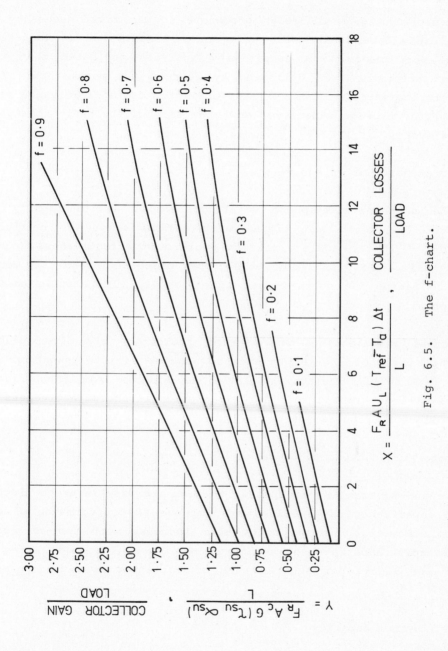

Fig. 6.5.   The f-chart.

and V is the collector air flow rate in litres/second per square metre of collector area.

To develop the general design procedure it was first necessary to study the effects that various design parameters had on long-term system performance. A sensitivity analysis indicated that the system performance was dominated by four system parameters: the collector orientation and tilt, the collector fluid capacitance rate $(m_c c_c)$, the storage capacity and the size of the heat exchangers in the system.

Collector orientation and tilt

In the southern hemisphere solar collectors should ideally be oriented due north and tilted at an angle equal to the local latitude + 15° to achieve maximum year-round effectiveness. For Sydney (33°, 52'S, 151° 12'E), the proper orientation and tilt would then be facing due north tilted approximately 49° relative to the horizontal.

In actual installations where solar collectors are often placed flush on the roof, it is not easy to acquire ideal orientation and tilt. Actually, relatively large deviations from these ideal angles will not result in a large reduction in the effectiveness of the collector (Figures 6.6 and 6.7).

Collector fluid capacitance rate

Duffie and Beckman [9] have shown that an increase in the collector fluid capacitance rate will always improve the performance of the system. However, pumping costs also increase with the collector fluid capacitance rate, so a compromise figure of around 0.8 kg of water per minute per square metre of collector area is normally used.

Storage system capacity

As was indicated earlier, the storage capacity of the system does have a significant influence on long-term system performance.

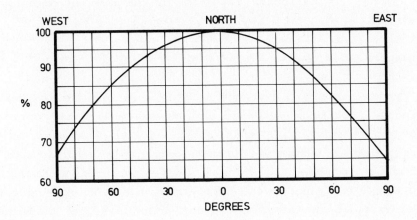

Fig. 6.6. Effect of collector orientation
           on annual heating performance.

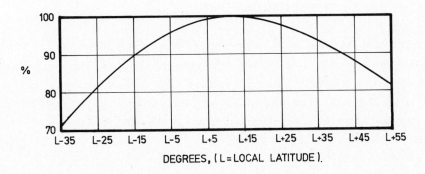

Fig. 6.7. Effect of collector tilt
           on annual heating performance.

However, an economic analysis by Lof and Tybout [1] (see Figure 6.1), showed that storage systems with between 50 and 75 kg of water per square metre of collector resulted in the most economic solar heating system.

Heat exchangers

Several types of heat exchangers are used in solar heating applications. For example, when anti-freeze solutions, corrosion inhibitors or fluids other than potable water are used in the collectors a heat exchanger is needed to transfer the heat from the collector fluid to the domestic hot water without allowing the fluids to mix. To transfer the heat across the heat exchanger from the collector fluid to the domestic hot water, a higher temperature must be maintained in the collector loop than would have been needed if the heat exchanger had not been there. This increases the collector temperature which causes the collector efficiency to decrease. So, to deliver the same useful energy, it is necessary to increase the area of the collector. As a rule of thumb, the collector area should be increased by about 2% for every $1^{\circ}C$ temperature drop across the heat exchanger. Since the temperature difference across a well designed heat exchanger is usually between 5 and $8^{\circ}C$, the addition of a heat exchanger will require an increase in the collector area of between 10 and 16%.

Application of the f-Chart Procedure

The f-chart in Figure 6.5 correlates the parameters X, Y and f, where X is the monthly heat loss from the collector per unit heating load, Y is the monthly energy gained by the collector per unit heating load and f is the fraction of the total load provided by the solar heating system for that month. In Chapter 2 the 160 m^2 residence shown in Figure 2.1 and located in Sydney was analysed to determine its energy requirements. This same structure will be used here to illustrate the use of the f-chart. That structure had a design heat load of 12.7 kW at a design temperature difference of $14^{\circ}C$. Consequently, the heat loss from the house was

12.7 kW (24 hr/day)/14°C   =   21.8 kWh/Degree Day

This means that for every degree Celsius that the average daily outside temperature is below 22°C, the house will need 21.8 kWh of heat for that day.   Some of this heat will be provided by internal sources, such as structural storage, lights, people, cooking, etc.

Table 4.3 gives the efficiencies measured on flat-plate solar collectors that are similar to the types currently sold in Australia.   Collector No. 4 will be used in this analysis.   This collector consists of several parallel copper tubes soldered to a copper absorber plate.   A black-chrome selective surface is used on the absorber and the collector is covered with one sheet of glass. The equation for the efficiency of this collector is:

$$\eta = 80 - 570 \, (T_i - T_a)/G_T$$

From this efficiency equation,

$$F_R U_L = 5.7 \text{ W/m}^2 \, ^\circ C$$
and

$$F_R \, \tau_{su} \, \alpha_{su} = 0.80$$

In this analysis north-facing collectors with three different angles of tilt will be considered (50°, 34° and 17°).   The system has 80 kg of water in the thermal storage system per square metre of collector area.

The monthly values for the f-chart parameters for this home are outlined in Table 6.3.   Part (a) shows the measured system parameters for this Sydney home taken from Chapter 4 and part (b) gives the actual form in which these parameters are used in the f-chart of Figure 6.5.   The average daily insolation (G_T) was taken from Table 3.4 and the degree-day data (Column 2) from Table 2.12. The monthly average ambient temperature (Column 3) was determined from the equation

TABLE 6.3   Monthly Values of Parameters Used in f-Chart Analysis

(a)   System Parameters from Chapter 4

	1. G_T MJ/m^2 - day Angle of Tilt			2. °Days per Month	3. T_a °C	4. $T_{Ref}-T_a$ °C	5. L_H MJ/month	6. L MJ/month
	50°	34°	17°					
Jan.	17.0	19.6	21.2	0	18.3	81.7	0	921
Feb.	16.3	18.1	19.0	0	18.3	81.7	0	923
Mar.	16.2	17.0	17.0	0	18.3	81.7	0	951
Apr.	16.6	16.5	15.4	0	18.3	81.7	0	1024
May	15.7	15.0	13.2	85	15.5	84.5	6681	7794
June	12.9	12.1	10.4	144	13.5	86.5	11318	12495
Jul.	16.4	15.4	13.3	170	12.6	87.4	13362	14567
Aug.	15.8	15.5	14.2	152	13.4	86.6	11947	13116
Sep.	18.1	18.7	18.1	81	15.6	84.4	6367	7480
Oct.	17.0	18.5	19.1	12	17.9	82.1	943	1986
Nov.	18.2	20.6	22.2	0	18.3	81.7	0	993
Dec.	18.6	21.5	23.5	0	18.3	81.7	0	946
						Total	63196	

(b)   f-Chart Parameters for Collector No.4 of Table 4.3

	X For Collector Area m^2			y/A_c,m^{-2} 50° Tilt	Y and Monthly f Angle of Tilt = 50° Collector Area					
					20 m^2		30 m^2		40 m^2	
	20	30	40	50°	Y	f	Y	f	Y	f
Jan.	27.08	40.62	54.16	0.458	9.16	1	13.73	1	18.31	1
Feb.	27.02	40.53	54.05	0.396	7.91	1	11.87	1	15.82	1
Mar.	26.23	39.34	52.46	0.423	8.45	1	12.67	1	16.90	1
Apr.	24.36	36.54	48.72	0.389	7.78	1	11.67	1	15.56	1
May	3.31	4.96	6.62	0.050	1.000	0.61	1.500	0.79	2.000	0.90
June	2.11	3.17	4.23	0.025	0.496	0.32	0.743	0.45	0.991	0.56
July	1.83	2.74	3.65	0.028	0.558	0.39	0.838	0.54	1.117	0.66
Aug.	2.02	3.02	4.03	0.030	0.598	0.41	0.896	0.56	1.195	0.68
Sept.	3.44	5.17	6.89	0.058	1.161	0.70	1.742	0.87	2.323	0.98
Oct.	12.62	18.93	25.24	0.212	4.246	1	6.369	1	8.491	1
Nov.	25.11	37.68	50.23	0.440	8.798	1	13.20	1	17.60	1
Dec.	26.37	39.55	52.7	0.488	9.752	1	14.63	1	19.50	1

Degree Days = $(18.3 - T_a)$(days per month)   (6.3)

The values in column 4 were obtained by subtracting T_a (column 3) from the reference temperature used in the f-chart (i.e. $100^{\circ}C$). Finally, the monthly heating load (L_H, column 5) is the product of the design heating load for the house (21.8 kWh per degree day) multiplied by the number of degree days in that month (column 2). The monthly heating load (L_H) and the total heating load (L) are obtained from Table 2.12.

Part (b) of Table 6.3 gives the X and Y factors used to find the monthly solar fraction (f) from Figure 6.5 for the three collector angles. The minimum slope for a tile roof in Sydney is approximately 17°, so 17° was used as one of the tilt angles. The 50° tilt is close to the angle recommended in Table 6.2 for Sydney (latitude plus 15°). However, a 50° roof angle may be architecturally unacceptable, so a 34° tilt is also included which is approximately half-way between the other two tilt angles. The first three columns of part (b) of the table give the X-parameters for three different collector areas (20, 30 and 40 square metres). This X-parameter is independent of the collector tilt angle. The last six columns of this table give the Y-parameter and the monthly solar fraction (f) for these same areas when the collector is mounted at the optimum tilt angle of 50°. The monthly solar fractions obtained from this analysis are plotted in Figure 6.8 for each of the three collector areas. The figure shows that the fraction of the heating load provided by the solar unit will vary significantly throughout the year. During the summer months, when the insolation is at its maximum and the only demand is from the domestic hot water system, the solar heating system will provide more hot water than is needed. However, during the winter months the demand is much greater and the insolation less, so much of the load must be carried by the auxiliary heating system. The annual proportion of the energy requirements provided by solar is calculated by summing the monthly energy provided by solar energy (i.e. the sum of $f \times G_T$ for each month) and dividing that by the total annual heating requirements (i.e. 63196 MJ from Table 6.3a). For collectors mounted at 50° (the angle recommended for Sydney) the annual solar fraction calculated for collector areas of 40, 30 and

20 square metres were 0.75, 0.65 and 0.52, respectively.   Since system costs are closely related to the collector area, an economic analysis will be required to determine which of these systems is actually best.   An economic analysis will be performed on this system in Chapter 11.

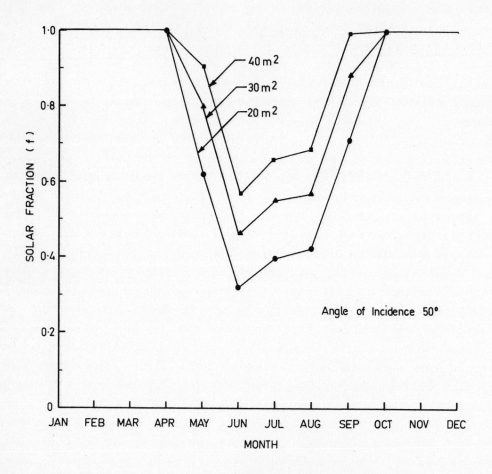

Fig. 6.8.   Variation in solar heating system performance throughout the year.

The tilt of the collector also has an effect on the energy collected each month.   In theory, the optimum tilt angle for solar collectors is the latitude plus 15°; however, local weather patterns can significantly affect the optimum tilt angle.   For example, if the winter months are normally mild and cloudy, but the autumn and spring periods normally clear and cool, collectors tilted at angles significantly less than the local latitude plus 15° may give a larger solar fraction than collectors mounted at the theoretical optimum angle.   The system considered here was analysed for tilt angles of 17°, 34° (the latitude of Sydney) and 50° (local latitude plus 16°).   Figure 6.9 shows that the insolation (G_T) on the 50° surface would be greater than on the other two surfaces for the coolest months (April through August).   However, the analysis found that for Sydney the annual solar fraction was only 2.8% greater for the 50° than for the 34° surface and only about 11% greater than for the 17° surface.   Therefore, in many installations, the collectors can be mounted directly on the existing roof rafters and a special supporting structure will not be needed.   This approach could save the cost of the special structure and reduce substantially the number of roofing tiles required.   It may also produce a more attractive structure.   A detailed economic analysis would be needed to select the best system for this application.

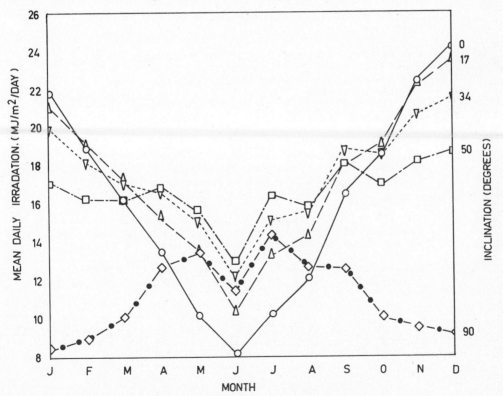

Fig. 6.9. Mean daily irradiation on an inclined surface in Sydney.

NOMENCLATURE

Symbol	Meaning

A_c — the total collector area (m^2)

f — the fraction of the total heating load provided by the solar-heating system during the time period

F_R — the "heat recovery factor" (dimensionless)

G_T — the instantaneous solar radiation incident on the collector (W/m^2)

L — total space and water heating loads for each calendar month (kJ)

m_T — the storage capacity of the tank (kg of water/m^2 of collector area)

q_u — the useful energy delivered by the collectors (W)

t — number of hours in each calendar month (hr)

t_s — the turnover time for the water in the storage tank (hr)

T_a — the ambient temperature of the atmosphere (^oC)

T_i — the temperature of the coolant at the inlet of the collector

T_p — the average temperature of the absorber plate (^oC)

T_{ref} — reference temperature (100^oC)

U_L — the overall heat-transfer coefficient for the heat lost from the absorber to the surroundings ($W/m^2 \ ^oC$)

α_{su} — the solar absorptivity of the absorber, i.e. the fraction of the solar energy reaching the absorber that is actually absorbed, (dimensionless)

τ_{su} — the solar transmissivity of the cover, i.e. the fraction of the solar radiation that reaches the absorber, (dimensionless)

REFERENCES

1. Lof, G.O.G. and Tybout, R.A., "Solar House Heating", Natural Resources Journal (April 1970).

2. Colorado State University, "Solar Heating and Cooling of Residential Buildings: Design of Systems", U.S. Government Printing Office, Washington, D.C. (1977).

3. Ward, J.C. and Lof, G.O.G., "Long-term (18 years) Performance of a Residential Solar Heating System", vol. 18, Solar Energy (1976) pp.301-308.

4. Telkes, M. and Raymond, E., "Storage Solar Heat in Chemicals - A Report on the Dover House", Heat and Vent., Vol.46 (Nov. 1949) pp.74-86.

5. Boehm, R.F. and Swanson, S.R., "Solar System Design Handbook for Utah", Pub. by : Utah Eng. Exp. Stn., Univ. of Utah (June 1978).

6. Morrison, D.J. and Abdel-Khalik, S.I., "Effect of Phase-Change Energy Storage Units on the Performance of Air Based Solar Heating Systems", vol. 20, Solar Energy, (1978) pp.57-67.

7. Information gathered by E. Baker during visit to the Los Alamos Solar Test Laboratories in 1982.

8. Copeland, R.J. et al., "The SERI Solar Energy Storage Program", Annual Thermal Storage Program Review, Tyson's Corner, Virginia (Dec. 1979).

9. Klein, S., Beckman, W. and Duffie, J. "A Design Procedure for Solar Heating Systems", Solar Energy, Vol.18, No.2 (1976).

10. Duffie, J. and Beckman, W. "Solar Energy Thermal Processes", New York, John Wiley and Son (1974).

11. Morrison, D.J., Sapsford, C.M. and Litvak, A., "Solar Insolation Data for Sydney", University of N.S.W., School of Mechanical and Industrial Engineering, Report No. 1979/FTM/2.

CHAPTER 7

Solar Cooling of Residential Buildings

INTRODUCTION

In any realistic evaluation of a solar energy assisted air conditioning system it is necessary to determine the cooling load for the building on an hourly basis for the period of the year in which air conditioning is required. This analysis requires information on the following, preferably on an hourly basis.

1. rate of solar insolation (beam and diffuse),
2. ambient temperature, wet bulb and dry bulb,
3. wind velocities and direction,
4. heat loads generated inside the structure.

From these data it is then possible to calculate the cooling load for a particular building at hourly intervals. Using the performance data for the selected solar collector and the cooling load, the appropriate collector area, the capacity of the storage system and the air conditioning equipment can be selected. In the next section the cooling load calculations for a residential building will be done manually using tables and charts from references [1] and [2]. A complete cooling load analysis of a system on an annual basis would require extensive computing time, particularly if system costs are to be determined. There are a number of proprietary programs available for determining air conditioning cooling loads, such as TEMPER and CAMEL, but which require mainframe computers to run. Programs for personal computers, such as Residential HVAC Loads [3] and CIRA [4] have begun to appear, but these are still quite expensive and they often need to be adapted to local conditions. A fundamental understanding of the basic heat-transfer theory is needed to be able to adapt these programs to local conditions. The purpose of the example used in this chapter is to demonstrate these heat transfer fundamentals.

COOLING LOAD CALCULATIONS

Before commencing the cooling load calculations it is first
necessary to evaluate the thermal aspects of the building.   The
thermal aspects of the building analysed in Chapter 2 (Figure 2.1)
are typical of the homes owned by families in the middle to upper
income bracket who would consider air conditioning as a necessary
requirement for, say, Sydney weather conditions.   Table 2.4
summarises the construction materials and heat transfer coefficients
for this building.

Before air conditioning is considered, the best possible indoor
conditions should be attained by the passive control functions of
the building.   Apart from the building structural, functional and
aesthetics requirements, the building should be thermally efficient.
Massive structure, minimum glass area, correct orientation of the
building, external shading devices for glass areas, plus adequate
usage of insulation materials will result in a thermally efficient
building.   However, other considerations (e.g. cost, asthetics,
etc.) may outweigh the thermal requirements and window areas may
exceed what is thermally desirable in such cases.

The structure chosen in this project is a composite of the above
requirements.   The walls have a mass of 195 kg/m^2, 100 mm thick
mineral fibre insulation with sarking is used in the roof and the
floor material is concrete (128 kg/m^2).   The building is correctly
orientated, there are no windows in the west facing wall, the window
area in the east-facing wall is at least 5 m^2 and the windows facing
north have an external shading device.   For the summer months direct
solar heat transmission is prevented by a horizontal shading device,
extending out 2.25 m above the 2.5 m high north-facing windows.
During the winter months, when the sun altitude angles are lower,
solar heat transmission will occur and help heat the space.

In the initial building design the area of roof facing north and the
slope must be considered.   An initial estimate of 80 m^2 of
collector area would necessitate a roof of approximately 100 m^2 at a
slope of say 20°.   For Sydney the sun altitude angle at noon on
January 21 is 76°, an optimum collector tilt angle for this

condition would be 14°. Considering winter operation of the collector system when the sun altitude angle is 45° (August 24) the optimum collector angle would be 45°. If all year round performance is required then a compromise angle would be 30°. In this study 20° is used (minimum angle for concrete tiles).

It is usual to consider 24 hour operation of the air conditioning plant for residential buildings and to size the plant capacity for 75% of the maximum cooling load. The cooling load is due to:-

1.   Solar heat transmission through the glass areas.
2.   Heat transmission due to a temperature difference through walls, roof, glass areas and floor.
3.   Water vapour flow through the structure.
4.   Heat due to air flow,
      i.e. infiltration and ventilation.
5.   Internal heat gain due to appliances.
6.   Heat from people.

Solar Heat Transmission Through Glass Areas

This solar heat transmission is due to both beam and diffuse insolation. The glass area, the type used, its shading and orientation all have a significant effect on the amount of solar heat transmitted to the conditioned space.

Consider as an example the portion of solar heat transmitted through a glass area for the following cases:-

1.   Reference glass* with a solar incident ray of 30°.
2.   Reference glass with a solar incident ray of 80°.
3.   Heat absorbing glass with a solar incident ray of 30°.

Figure 7.1 shows that some of the incident radiation (G_T) is reflected from the surface of the glass, some is absorbed by the glass and some is transmitted through the glass. The properties of

*Reference glass is defined in [1] as glass of 3 mm thickness and having a transmissivity (τ) of 0.86 and an absorptivity (α) of 0.06 for incident ray of 30°.

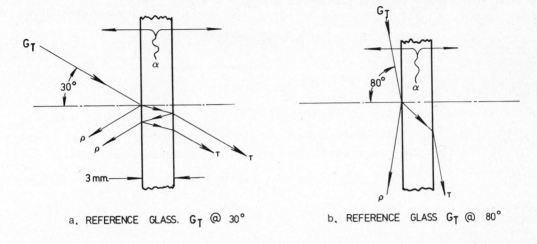

a. REFERENCE GLASS. G_T @ 30°   b. REFERENCE GLASS G_T @ 80°

Fig. 7.1 Effect of incidence angle on path length through glass.

the glass responsible for this dispersion are the reflectivity (ρ), the absorptivity (α) and the transmissivity (τ). These properties are related by,

$$1 = \alpha + \rho + \tau \tag{7.1}$$

Approximately 40% of the radiation absorbed by the glass will be transmitted to the conditioned space, so the heat transfer (q) into the space is,

$$q = G_T (\tau + 0.4\,\alpha) \tag{7.2}$$

Case 1

For reference glass

$$\tau = 0.86$$
$$\alpha = 0.06$$

The portion of G_T that enters the space (Figure 7.1a) is :

$$q = 0.86\ G_T\ +\ (0.4 \times 0.06\ G_T)$$

$$q = 0.884\ G_T$$

It is usual to refer to the fraction of the insolation (G_T) actually entering the conditioned space in terms of "glass factors" where $0.884\ G_T$ is taken as a glass factor of one, therefore in Case 1, the glass factor is one.

Case 2

Consider again reference glass but with the incident ray at an angle of $80°$ (Figure 7.1b). For these conditions the transmittance for the glass is 0.42 and the absorptance is 0.06. Thus, the portion of G_T that enters the space is :-

$$q = 0.42\ G_T + (0.4 \times 0.06\ G_T)\ =\ 0.444\ G_T$$

$$\text{Glass factor}\ =\ 0.444/0.884\ =\ 0.5$$

Case 3

Using heat absorbing glass at an incident angle of $30°$, where a is 0.52 and τ is 0.43, the portion of G_T that enters the space is:

$$q = 0.43\ G_T + (0.4 \times 0.52\ G_T)\ =\ 0.643\ G_T$$

$$\text{Glass factor}\ =\ 0.643/0.884\ =\ 0.73$$
(Less heat is transmitted than in Case 1).

Table 7.1 gives solar characteristics and glass factors for a range of glasses.

Table 7.2 [2] shows the solar heat gain through reference glass (W/m^2) for exposure and time of day for a South latitude of $34°$. These are calculated values and allowances should be made for cloud, haze and smog.

TABLE 7.1   Solar Characteristics and Glass Factors for Various Glasses [2]

	Type of Glass	Absorptivity (α)	Reflectivity (ρ)	Transmissibility (τ)	Glass Factor
C A R R I E R	Reference (3 m clear)	0.06	0.08	0.86	1.00
	Plate clear (6 mm)	0.15	0.08	0.77	0.94
P I L K I N G T O N	"Spectrafloat" 50/67 Bronze (6 mm)	0.34	0.10	0.56	0.80
	"Antisun" float 41/62 Grey (6 mm)	0.51	0.05	0.44	0.73
	"Antisun" float 29/46 Bronze (12 mm)	0.74	0.05	0.21	0.58
	"Calorex" 48/45 Blue/Green (6 mm)	0.75	0.05	0.20	0.57
C S I R O	Laminated "Stopray" Gold (6 mm)	0.36	0.39	0.25	0.44
	"Reflectoshield" control film R.S.20 on clear glass (6 mm)	0.44	0.44	0.12	0.34
	"Scotchtint" control film A18 on clear glass (4 mm)	0.30	0.53	0.17	0.33

TABLE 7.2   Solar Heat Gain Through Reference Glass (W/m^2)   [2]

		Time						
Exposure		06.00	08.00	10.00	12.00	14.00	16.00	18.00
34° South Latitude (Jan.)	North	15	42	109	165	109	42	15
	South	38	43	47	47	47	43	77
	East	380	550	330	47	47	40	15
	West	15	40	47	47	330	550	370

Using the window areas (Figure 2.1) and the heat gains from Table 7.2 it is now possible to calculate the solar heat gain (q_s) to the space where:

$$q_s = A \times G_T \times \tau \times \text{glass shade factor} \times \text{drape}$$
$$\text{shade factor} \times \text{storage factor} \qquad (7.3)$$

where q_s =   cooling load (watts)

A   =   window area (m^2)

G_T =   Insolation from Table 7.2 (W/m^2)

=   0.86

glass shade factor   =   0.95, from tables [2]

drape shade factors   =   0.7, from tables [2]

The storage factor is obtained from Table 7.3.

TABLE 7.3   Storage Load Factors, Solar Heat Gain Through Glass with Internal
             Shading Device [2]

Mass Per Unit Area of Floor	Time / Aspect	Constant Space Temperatures						
		06.00	08.00	10.00	12.00	14.00	16.00	18.00
375 kg/m^2	N	0.14	0.37	0.68	0.82	0.64	0.34	0.16
	S	0.98	0.98	0.98	0.98	0.98	0.98	0.98
	E	0.52	0.78	0.62	0.27	0.20	0.15	0.12

Effects of Storage

Heat gained in an air conditioned space becomes part of the cooling load (i.e., the heat that must be removed by the air-conditioning equipment).   In the case of radiant heat such as that coming from the sun, lights and even people, the radiant energy is first absorbed and stored in the surfaces that it strikes.   This raises the temperatures of the surfaces above the surroundings and the stored heat is eventually transferred to the air by convection and

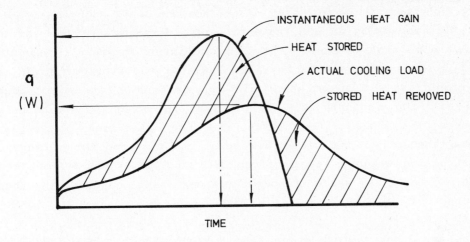

q
(W)

INSTANTANEOUS HEAT GAIN

HEAT STORED

ACTUAL COOLING LOAD

STORED HEAT REMOVED

TIME

(a)   Comparison of cooling load with
the instantaneous heat gain.

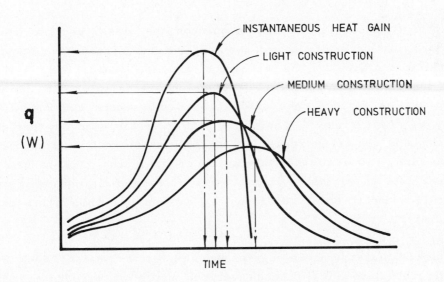

q
(W)

INSTANTANEOUS HEAT GAIN

LIGHT CONSTRUCTION

MEDIUM CONSTRUCTION

HEAVY CONSTRUCTION

TIME

(b)   Comparison of the cooling load
and the type of construction.

Fig. 7.2. Some factors that influence the hourly cooling load.

only then does it become part of the cooling load.   Obviously, all radiant heat gains do not instantaneously form part of the cooling load and a "storage factor" is used to account for this in equation (7.3).   It can be seen from Table 7.3 that the storage factor can be significant in solar heat gain calculations.   It is not usual for all the instantaneous heat gains to occur simultaneously. Figures 7.2a and 7.2b illustrate the relationships between instantaneous heat gain and the actual cooling load for a range of construction weights ranging from 150 to 700 kg per m^2 of floor area.   Constant space temperature has been assumed in these figures.

There are other factors which tend to reduce the cooling load, such as:

1.   Diversity factors for people, lights, appliances.
2.   Shading (trees, landscape, structures, etc.).

Heat Transmission due to Temperature Difference

In addition to the solar heat gain through the glass, heat is also transmitted to the space due to the temperature differences between the outside surface of the structure and the inside air.   The outside surface temperatures are a function of the solar heat being absorbed by the surface, the ambient air temperature and the wind velocity at the surface.   These are highly variable at any period of the day and generally result in unsteady state heat flow through the structure.   Due to the difficulty of analysing this unsteady state heat flow an equivalent temperature difference is used whereby the heat flow due to solar radiation and ambient air temperature can be calculated.

The equivalent temperature difference (ΔT_e) is established by measuring the heat flow through various types of structures for variable solar radiation and ambient air temperatures, building latitude, time of day and design conditions.   Tables 7.4 and 7.5 give equivalent temperature differences for a design day in January at 34° south latitude for walls and roofs respectively.

TABLE 7.4   Equivalent Temperature Differences for Walls (°C) [2]

Time Exposure	06.00	08.00	10.00	12.00	14.00	16.00	18.00	20.00	22.00	24.00	02.00	04.00
NORTH	1.7	0.0	1.1	8.9	15.6	16.7	13.3	8.9	6.7	4.4	2.8	2.2
SOUTH	0.6	0.0	1.1	2.2	5.6	7.8	8.9	8.9	6.7	4.4	2.8	1.7
EAST	1.7	2.2	18.9	19.4	10.0	8.9	8.9	8.9	7.8	5.0	3.9	2.8
WEST	3.3	2.2	2.2	4.4	7.8	16.7	24.4	22.2	11.1	5.6	4.4	3.9

TABLE 7.5   Equivalent Temperature Differences for Dark Coloured Roofs (°C) [2]

Time Exposure	06.00	08.00	10.00	12.00	14.00	16.00	18.00	20.00	22.00	24.00	02.00	04.00
Sunlit	2.2	1.1	3.3	11.1	18.9	25.0	26.0	21.7	16.1	10.6	6.7	4.4
Shaded	-0.6	0.0	2.2	5.6	8.9	10.0	8.9	6.7	3.3	2.2	0.6	-0.6

The heat transmitted into the space (q_t) can be calculated from:

$$q_t = U \, A \, \Delta T_e$$

where ΔT_e is the equivalent temperature difference ($^\circ$C) from Tables 7.4 and 7.5.

Water Vapour Gain Through the Structure

The water vapour flow through the structure is essentially a latent heat load, and is due to vapour pressure differences between the outside and inside air.   To prevent vapour condensation within the walls, vapour barriers are usually applied.   The latent heat load in residential buildings due to vapour flow through the structure is usually insignificant and will not be considered in this application.

Heat Gain Due to Infiltration and Ventilation

Infiltration and ventilation introduce outside air into the conditioned space which increases the sensible and latent heat gains.

For this building a ventilation rate of 5 litres/s per person will be used, therefore for 4 persons this represents 0.2 air changes per hour.   Using a conservative approach ventilation and infiltration will be assumed to total 0.5 changes/hr. Since the building volume is 400 m^3, this represents a flow rate of 55.5 litres/s.

The sensible heat load (q') is:

$$q' = m \, c_p \, \Delta T = 1.2 \, Q_{OA} \, \Delta T \qquad\qquad (7.4)$$

where

ΔT = (outside air temperature - room design temperature)

The latent heat load (q) is

$$q = 3 \times Q_{OA} (W_{OA} - W_{RA}) \qquad (7.5)$$

where W_{OA} and W_{RA} are the moisture content of the outside air and the room air (g/kg), respectively, and Q_{OA} is the volume of outside air entering the space in litres per second.

Internal Heat Gains

Typical internal heat gains from appliances (sensible and latent) are given in Table 7.6 with peaks occurring at 0600, 1400 and 1800 hours.   In addition, an internal heat gain of 100 W will be used for each person (67 W sensible, 33 W latent).   The heat loads calculated at 2 hourly intervals is given in Table 7.6 and plotted in Figure 7.3.   The peak load occurs between 1800 and 2000 hours.

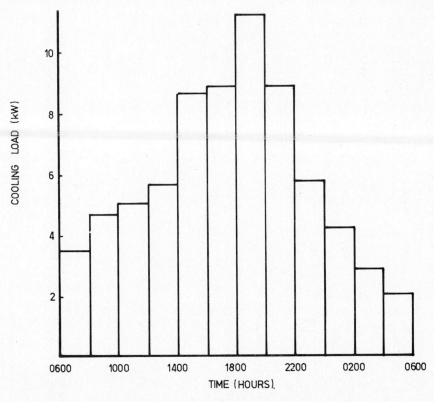

Fig. 7.3 Cooling load profile for January 21.

TABLE 7.6   January Cooling Load Estimate

Structure or Heat Path	Area m²	Aspect	06.00	08.00	10.00	12.00	14.00	16.00	18.00	20.00	22.00	24.00	02.00	04.00
Glass														
Solar Heat Transmission $q = A \times G_T \times \tau_{su} \times .95 \times .7 \times S.F.$	35	N	106	318	639	771	602	320	247	–	–	–	–	–
	20	S	426	482	527	527	527	527	860	–	–	–	–	–
	5	E	565	1226	585	36	36	17	5	–	–	–	–	–
Σ1.			1097	2028	1751	1334	1165	864	1112	–	–	–	–	–
Walls & Glass	60	glass	–	212	389	778	1979	2756	3145	3145	2367	1554	989	600
Heat Transmission $q = AU\Delta T$	15	N	–	52	34	270	475	509	405	271	204	134	85	67
	30	S	–	36	67	134	341	475	542	542	408	268	170	104
	15	E	52	67	575	591	305	271	271	271	238	152	119	85
	30	W	134	89	89	178	317	678	990	901	450	227	179	158
Σ2.			186	456	1154	1951	3417	4689	5353	5130	3667	2335	1542	1014
Roof	100	N	80	40	120	405	689	912	949	792	587	387	244	160
	74	S	–	–	80	204	325	365	325	245	120	80	22	–
Σ3.			80	40	200	609	1014	1277	1274	1037	707	467	266	160
Infiltration & Ventilation	55 ℓ/s		–	33	132	231	363	363	330	198	99	33	–	–
People (4)			268	268	268	268	268	268	268	268	268	268	268	268
Appliances			900	600	300	300	900	300	2000	1000	300	300	200	200
Σ4.			1168	901	700	799	1531	931	2598	1466	667	601	468	468

TABLE 7.6 (Continued)   January Cooling Load Estimate

		06.00	08.00	10.00	12.00	14.00	16.00	18.00	Time 20.00	22.00	24.00	02.00	04.00
Sensible Heat Load Σ 1.2.3.4.	Σ5.	2531	3425	3805	4693	7127	7761	10337	7633	5041	3403	2276	1642
Latent Heat Load													
People (4)		132	132	132	132	132	132	132	132	132	132	132	132
Infiltration & Ventilation		500	759	710	710	842	842	627	759	710	759	500	500
Appliances		300	300	–	–	300	–	300	300	–	–	–	–
	Σ6.	932	1191	842	842	1274	974	1059	1191	842	891	632	632
Total Heat Load Σ5.6	Σ7.	3463	4616	4647	5535	8401	8735	11396	8824	5883	4294	2908	2274
E.S.H.F. = Σ5/Σ7		0.73	0.74	0.82	0.85	0.85	0.88	0.9	0.86	0.85	0.79	0.78	0.72
Cooling Load MJ/hr (Average)		12.5	16.6	16.7	19.9	30.2	31.5	41.0	31.8	21.2	15.4	10.4	9.3
Daily Cooling Load	(January = 513 MJ)												

COOLING EQUIPMENT

The results from the cooling load analysis on the residence in
Figure 2.1 will be used to size and select an appropriate solar
cooling system for that building.   Ideally, hourly data for both
the cooling load and for the insolation would be used in this
analysis and a computer simulation would be performed for the system
to optimise the collector area and the storage requirements.
Computer programs are available which determine the variations in
the solar fraction for the cooling period.   This section will
introduce the fundamental equations used for these simulation
studies to provide insight into the criteria used for equipment
sizing.

SELECTION OF A SOLAR ENERGY COOLING SYSTEM

A lithium bromide-water ($Li.Br.H_2O$) absorption chiller will be
considered for the cooling calculations.   The system will use flat
plate collectors with selective surfaces as well as both hot water
and cold water storage.   The hot water storage will limit the
transient heat fluxes to the absorption unit, the cold storage will
prevent cycling of the absorption unit and in conjunction with the
hot storage, provide cooling when there is insufficient solar
insolation.

Figure 7.4 shows the basic components of a $Li.Br.H_2O$ water chiller
typical of those used for solar energy operation within the cooling
capacity range of 7 to 24 kW.   In comparison to a vapour
compression system the absorber performs the equivalent of the
suction stroke portion of the compressor and the generator the
discharge stroke.   The condenser and evaporator serve the same
functions as in the vapour compression system.

In this system the water (refrigerant) vapour flows into the
absorber where it is absorbed by the lithium bromide solution
(absorbent).   To maintain the temperature and hence the pressure in
the absorber needed to draw the water vapour from the evaporator,

cooling water (or air) is used to remove the heat released during the absorption process.   The absorbent rich with refrigerant, often referred to as the "weak absorbent solution" *, is transferred to the generator where much of the refrigerant is evaporated by the application of heat.   With much of the refrigerant removed, the "strong absorbent solution" flows back to the absorber through a pressure reducer where again it will be enriched with refrigerant to start the cycle again.   While the refrigerant circulates through the entire system, the absorbent is confined to the absorber-generator circuit.

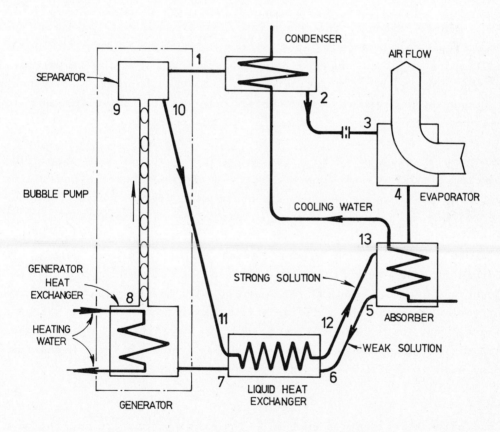

Fig. 7.4 Li.Br. absorption air conditioner with a bubble pump.

*By definition the "weak absorbent" solution is that which picked up refrigerant in the absorber and is now weak in its affinity for refrigerant. A "strong absorbent" solution is one from which the refrigerant has been extracted and has a strong affinity for refrigerant.

The system in Figure 7.4 uses a liquid-to-liquid heat exchanger to remove heat from the hot stream of strong absorbent leaving the generator and to transfer this heat to the stream of weak absorbent solution entering the generator.

The temperature of the condenser cooling medium will determine the pressure in the generator and the condenser. For example, at a condensing temperature of $38^{\circ}C$ the pressure is 50 mm Hg (abs). Similarly, the pressure in the evaporator and the absorber is established by the evaporating temperature.

To a large extent, the generator performance will dictate the economical performance of the system. Hot water from the solar collectors supplies heat to the generator to raise the temperature of the weak absorbent solution coming from the absorber. This heat is needed to evaporate the water out of the solution, and to overcome losses that are incidental to the process.

The pump in Figure 7.4 is a thermally driven vapour lift pump. The pumping method relies upon a density difference between the fluids entering and leaving the high pressure side of the system. Vapour boiled out of solution in the generator heat exchanger collects at the base of the pump tube with slugs of strong absorbent. This mixture of liquid and vapour having a low effective density is forced up the pump tube by the heavier solution entering the generator. This method of pumping liquid is limited to low differential heads and relatively low flow rates. Its use has been limited to $Li.Br.H_2O$ chillers of low capacities. For higher capacities this thermal pump is usually replaced by a mechanical pump such as that shown in the system in Figure 7.5.

Figure 7.6 shows the thermodynamic cycle for the $Li.Br.H_2O$ absorption refrigeration system. The state points are plotted on a pressure, temperature, concentration diagram where the key cycle state points are indicated as solid lines while the dotted lines represent the system temperatures which establish the operational limits.

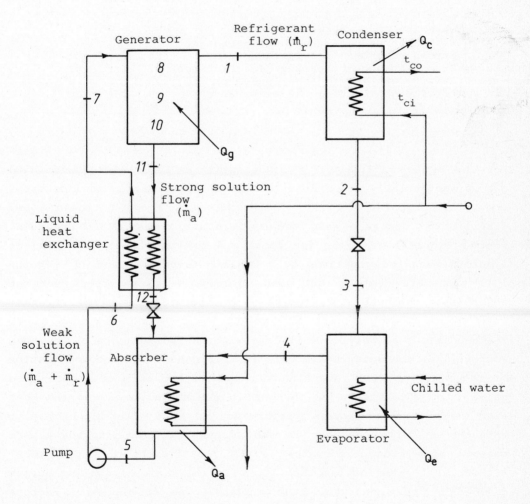

Fig. 7.5. Lithium-bromide-water absorption refrigeration cycle.

For an evaporation temperature of 4°C, a condensation temperature of 35°C, and a mean generator temperature of 88°C, the cycle state points are:

State Point	SOLUTION	Solution Temp., $^{\circ}$C	Properties % (Li.Br.)
5	Weak absorbent leaving absorber	32	56
7	Weak absorbent leaving heat exchanger	71	56
11	Strong absorbent leaving generator	85	61
12	Strong absorbent leaving heat exchanger	42	61
2	Refrigerant leaving the condenser	38	0
4	Refrigerant leaving the evaporator	4	0

For the cycle in Figure 7.6 condensation of the refrigerant (H_2O) occurs isothermally at point 2 (38°C), and evaporation is isothermal occurring at point 4 (4°). Between points 2 and 1, boiling and distillation occurs within the temperature range of 72° to 85°C. In the liquid heat exchanger (points 6 to 7 and 11 to 12) the hot liquid leaves the generator at a maximum temperature of 85°C (point 10), transfers heat to the weak solution being returned from the absorber to the generator, (points 6 to 7).

The conditions at point 1 will determine the temperature of the heat source (i.e. the solar collectors in this case). The concentration of the solution will determine the temperature at both point 1 and point 12. The effect of changes in the heat source and the heat sink temperatures on cycle performance is best determined by heat balances on each component in the cycle. The most commonly used measure of system performance is the "coefficient of performance" (C.O.P.). By definition the C.O.P. is the ratio of cooling output divided by the generator heat input. One of the many factors influencing the C.O.P. is the relative recirculation rate which is defined as the ratio of the mass flow rate of the strong absorbent to the refrigerant mass flow rate. High recirculation rates require more pump work and therefore reduce the C.O.P.

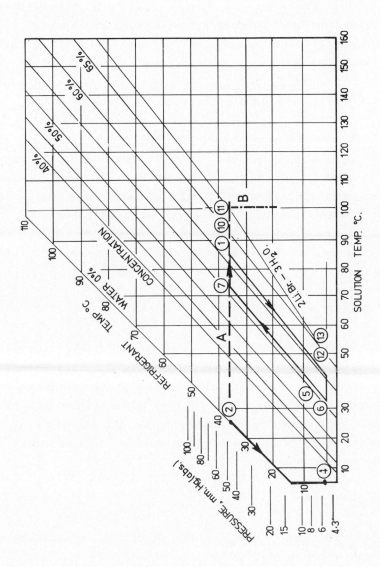

Path A   Ideal condensing pressure.

Path B   Maximum generator temperature.

Fig. 7.6.   Single-stage lithium-bromide-water cycle.

REFRIGERATION CYCLE SELECTION

There are a number of refrigeration cycles that can be considered as suitable for space cooling, when the heat source is solar energy. Some of these cycles are shown in Figure 7.7. In theory these cycles will operate from heat sources with temperatures ranging from 50 to 200°C. The pertinent parameters for such cycles are their "Heat Source" and "Heat Sink" limitations as a function of coefficient of performance (C.O.P.). The Dual Carnot is an ideal cycle with a C.O.P. of 1 at a supply temperature of about 90°C.

From Figure 7.7 it can be seen that the operating range of the Li.Br.H$_2$O cycle is compatible with the output of flat plate collectors. In contrast, the aqua-ammonia cycles operate efficiently only at temperatures that are too high for most flat plate collectors necessitating the use of evacuated-tube or concentrating collectors. Aqua-ammonia cycles do have the advantage that they can be air-cooled (which is not feasible with Li.Br.H$_2$O cycles).

AN EXAMPLE OF EQUIPMENT SELECTION

A Li.Br.H$_2$O water-cooled absorption chiller which has performance characteristics given in Table 7.7 will be used for this example. The C.O.P. for a range of generator inlet hot water temperatures and chilled water temperatures are shown in the table, hence the heat required to obtain the desired cooling output can be calculated. From a knowledge of the energy requirments it is then possible to size the collector area requirements.

COLLECTOR AREA

The collector area required to operate the absorption system can be determined from:

    a.   Daily insolation data, cooling loads and performance of a
          particular collector (in this case the Yazaki Blue Panel
          Model SC-201S will be used [5]).

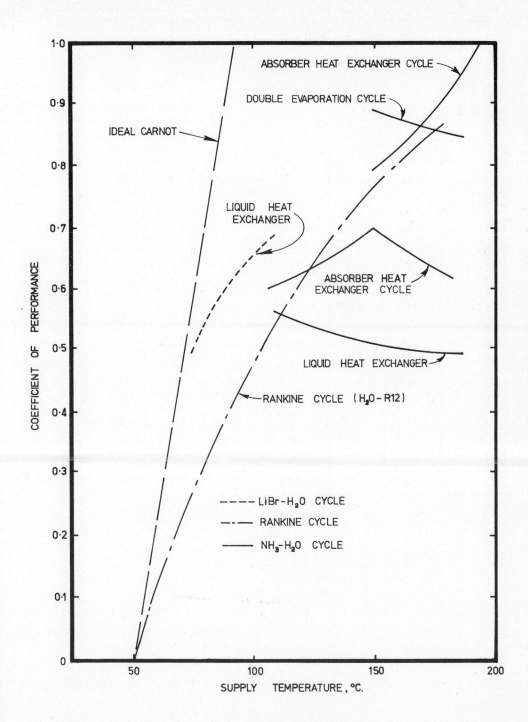

Figure 7.7.   Performance of heat-operated refrigeration cycles.

TABLE 7.7 Performance Characteristics of Two Commercial Chillers.
Lithium Bromide-Water Cycle Units Powered by Hot Water [5,6]

Yazaki Chiller Model WFC-600-S				Arkla Solaire Chiller Model WF.36			
Generator Hot Water Inlet Temp. °C	Chilled Water Temp. °C		C.O.P.	Generator Hot Water Inlet Temp. °C	Chilled Water Temp. °C		C.O.P.
	Inlet Temp.	Outlet Temp.			Inlet Temp.	Outlet Temp.	
70	14	11.0	0.40	*			
72	14	10.6	0.46	*			
74	14	10.1	0.50	*			
76	14	9.6	0.54	76	8.2	7.2	0.44
78	14	9.1	0.59	78	8.8	7.2	0.54
80	14	8.7	0.64	80	9.3	7.2	0.61
82	14	8.2	0.68	82	10.3	7.2	0.67
84	14	8.1	0.68	84	11.0	7.2	0.68
86	14	7.9	0.68	86	11.3	7.2	0.71
90	14	7.5	0.68	90	13.0	7.2	0.72
94	14	7.2	0.68	94	13.7	7.2	0.70
96	14	-	-	96	14.0	7.2	0.68

*
Permissible generator hot water inlet
temperature range (76°C to 96°C)

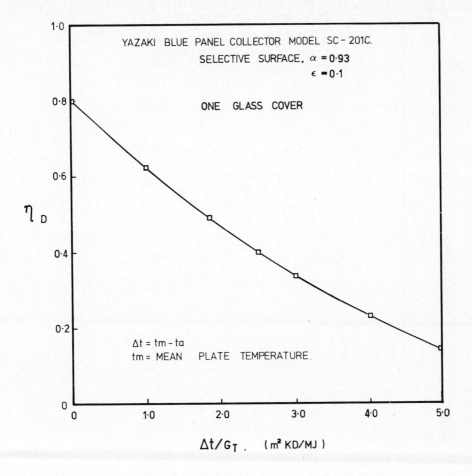

Fig. 7.8 Daily efficiency of the collector.

b. Instantaneous insolation data, hourly cooling loads and the instantaneous performance of the collector [7].

Figures 7.8 and 7.9 illustrate the daily and the instantaneous performance of the collector. Since the collector area is also a function of the C.O.P. of the absorption chiller, the variation in C.O.P. with respect to inlet hot water temperature to the generator and the temperature of the chilled water in the evaporator must also be considered. A constant condenser inlet water temperature will be assumed.

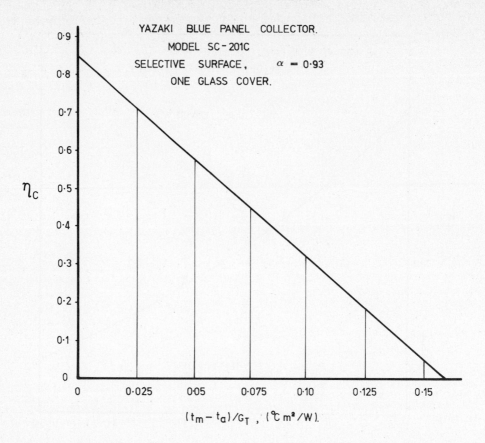

Fig. 7.9   Instantaneous collector efficiency.

Sizing of the Collector Area (METHOD A)

Although this method is not as accurate as Method B, it is more suitable for manual calculations.   Method A uses the following data:

G_T Daily solar insolation, MJ/m^2 per day (Table 7.8)

q_D Daily cooling load, MJ/day (Table 7.6)

η_D Collector daily efficiency

T_s Water temperature at start of heat collection, $^\circ C$ (Table 7.4)

T' Water temperature at end of heat collection, assuming no heat removed from storage (Table 7.9)

TABLE 7.8 Hourly Insolation and Temperature Data for January 21, Sydney

	Hour E.S.T.													
	0600	0700	0800	0900	1000	1100	1200	1300	1400	1500	1600	1700	1800	1900
G_T (MJ/m^2)	0.04	.4	1.18	2.11	2.7	3.1	3.52	3.56	3.35	2.94	2.37	1.67	.92	.24
					[Daily Total = 28.10 MJ m^{-2} D^{-1}]									
G_T (W/m^2)	11	117	327	586	750	861	978	989	931	817	658	464	256	67
T_a °C	16	19	19	23	28	29.2	30.4	31.3	31.7	31.3	30.1	28.4	26.9	26.0
					[Daily Mean = 29°C]									

TABLE 7.9   Collector Area Requirements on an Hourly Basis

	Hour E.S.T.									
Item	0800	0900	1000	1100	1200	1300	1400	1500	1600	1700
T_i (°C)	72.3	73.4	75.5	78.1	81.6	87.0	88.7	87.6	82.9	75.4
T_o (°C)	72.7	75.6	79.7	83.1	87.2	92.5	93.8	92.1	86.3	77.6
$\Delta T/G_T$	.14	.09	.07	.06	.058	.06	.066	.075	.087	.11
ηc (%)	10	37	46	51	54	51	47	45	39	25
C.O.P.	.46	.50	.64	.68	.68	.68	.68	.68	.68	.59
q_ℓ (Ave.) (MJ/hr)	16.6	16.6	16.7	18.3	19.9	25.0	30.2	30.8	31.5	36.3
q_g (MJ/hr)	36	33.2	26.1	26.9	29.7	36.7	44.4	45.3	46.3	61.5
G_T (MJ/m^{2}hr)	1.18	2.11	2.7	3.1	3.52	3.56	3.35	2.94	2.37	1.67
A_c (m^2)	339	47	23	19	17	23	26	38	56	164

T_a Mean ambient air temperature during daylight hours
($^\circ$C, Table 7.8).

H_L Percent of heat lost from the system.

Figure 7.8 shows the daily performance of the Yazaki Blue Panel collector [5].   The collector efficiency η_D can be obtained from this figure if the average plate temperature (T_p), the ambient temperature and the daily solar insolation are known, since:

$$T_p - T_a \approx [(T_s + 2T')/3 - T_a] \tag{7.7}$$

The collector area (A_c) can now be estimated from the relationship:

$$A_c = f\ q_D/[G_T\ \eta_D\ (1 - H_L)\ C.O.P.] \tag{7.8}$$

EXAMPLE 4 :   Method A

For the design day used in the cooling load example:

G_T = 28.10 MJ/m^2 per day, T_a = 29°C
T_s = 73.4°C,   T' = 88.7°C; q_{1D} = 513 MJ/D

and

$T_p - T_a$ = [(73.4 + 2 x 88.7)/3 - 29) = 54.6°C

From Figure 7.8, for ($T_p - T_a$) = 54.6°C

η_D = 43.5%

If,

f = 0.7

C.O.P. = 0.68

then the collector area is

A_c = (0.7 x 513)/[0.68 x 28.1 x 0.435 (1 - 0.1)]
A_c = 48 m^2.

Sizing of Collector Area (METHOD B)

Using hourly insolation data, temperature data and cooling load data together with the C.O.P. for the absorption chiller, Figure 7.9 can be used to estimate the collector area requirements at hourly intervals. The collector areas determined by this method will be different for each hour since the calculations are based on an instantaneous availability of solar energy and a varying cooling load.

$$A_c = f \times L/[G_T \ \eta_c(1 - H_L) \qquad\qquad (7.9)$$

Table 7.10 is a summary of the results used to calculate the collector areas where:

T_i = collector inlet water temperature, $^\circ C$

T_o = collector outlet water temperature, $^\circ C$

T_p = $(T_i + T_o)/2$

T_a = ambient temperature, $^\circ C$

G_T = solar insolation ($MJ/m^2 hr$) at a collector tilt
             angle of 20°

η_c = instantaneous collector efficiency (from Figure 7.9)

C.O.P. = chiller coefficient of performance
             (from Table 7.6 - Yazaki Chiller)

L   = hourly cooling load (MJ/h)   Table 7.5.

q_g = heat input to the generator (= L/C.O.P.)

The solar insolation and ambient temperature data used for this example are given in Table 7.8.

Assumptions have to be made regarding the collector inlet and outlet water temperatures. The values of T_i and T_o used in this example are the results of a computer simulation study [7].

Table 7.9 shows the collector area that would be needed to provide 100% of the hourly cooling load for each hour between 8 a.m. and 5 p.m. (Note that the area requirements at 8 a.m. and 5 p.m. in excess of the roof area.) Actually, these areas are not realistic as they assume that the solar fraction is 100%, and they do not take advantage of the savings offered by thermal storage.

Using collector areas of 40, 45 and 47 m^2 in equation (7.9) give solar fractions of 0.6, 0.68 and 0.71 respectively. These results are shown in Table 7.10.

A collector area of 47 m^2 will be used in this example (with a solar fraction of 0.71 at the design conditions). This solar fraction is based on a particular design day when the high cooling load also coincides with a day of high solar insolation. High cooling loads can also occur on days of high humidity and medium to low solar insolation levels. Under such conditions the solar fraction will be less than 0.71, so storage and/or auxiliary power would be a necessity if design conditions are to be met.

ENERGY DISTRIBUTION

A summary of the energy distribution for this example is shown in Table 7.11. The Yazaki chiller chosen has a maximum cooling capacity of 35 MJ/hr. With a peak of 0.68 (Table 7.7) the maximum useful energy input to this chiller is 51 MJ/hr.

If conditions at say 1200 hours are considered then the energy distribution may be arranged as follows:-

Cooling load (q_L) = 19.9 MJ/h
Collector system output (47 m^2) = 89.3 MJ/h
Chiller input requirements to match the cooling
 load (q_g) = q_L /C.O.P. = 29.6 MJ/h
Excess energy = 89.3 - 29.3 = 60 MJ

It is possible to store this excess energy either as hot water or partly as hot water and partly as chilled water. It is more economic to operate the chiller at maximum capacity (51 MJ/hr) throughout the day, in which case, since the load is only 19.9 MJ/hr, the system would produce the following at 1200 hours:

q_c = Energy output from the collectors = 89.3 MJ/h
q_g = Energy input to the chiller   = 51.5 MJ/h
q_{h-s} = Energy to hot storage   = 37.8 MJ/h
q_{c-s} = Energy to cold storage   = 15.1 MJ/h
q = Energy to match load   = 19.9 MJ/hr

TABLE 7.10   Collector Area Versus Solar Fraction for 24 Hour Operation

A_c = 40 m^2	Hours for Collector Efficiencies > 10%										Totals (MJ)
	0800	0900	1000	1100	1200	1300	1400	1500	1600	1700	
q_ℓ (MJ)	16.6	16.6	16.7	18.3	19.9	25.0	30.2	30.8	31.5	36.3	(513/Day)
q_c (MJ/hr)	4.72	31.2	49.7	63.2	76.0	72.6	63	53	37	16.7	(467)
q_R (11)	2.2	15.6	31.8	35	35	35	35	35	25.1	9.8	(260)
(q_R-q_ℓ) (MJ/hr)	12.4	1.0	15.1	16.7	15.1	10	4.8	4.2	6.4	26.5	
$q_{H.S}$ (MJ/hr)	-	-	-	11.7	24.5	21.1	11.5	1.5	-	-	(70.3)

Solar Fraction = (260 + 70.3 x 0.68)/513 = 0.60
(24 Hour Operation)

A_c = 45 m^2

	0800	0900	1000	1100	1200	1300	1400	1500	1600	1700	
q_c (MJ/hr)	5.31	35.1	35.9	71.1	85.5	81.7	70.9	59.6	41.6	18.8	(525)
q_R (MJ/hr)	2.4	17.6	35	35	35	35	35	35	28.3	11.1	(269)
(q_R-q_ℓ) (MJ/hr)	14.2	1.0	18.3	16.7	15.1	10	4.8	4.2	3.2	25.2	
$q_{H.S.}$ (MJ/hr)	-	-	4.5	19.6	34	30.2	19.4	8.1	-	-	(116)

Solar Fraction = (269 + (116 x .68)/513 = 0.68

A_c = 47 m^2

	0800	0900	1000	1100	1200	1300	1400	1500	1600	1700	
q_c (MJ/hr)	5.5	36.7	58.4	74.2	89.3	85.3	74.0	62.3	43.5	19.6	(549)
q_R (MJ/hr)	2.5	18.4	35	35	35	35	35	35	29.6	11.6	(272)
q_R-q_ℓ (MJ/hr)	14.1	1.8	18.3	16.7	15.1	10	4.8	4.2	1.9	24.7	
$q_{H.S.}$	-	-	6.9	22.7	37.8	33.8	22.5	10.8	-	-	(135)

Solar Fraction = (272 + (135 x 0.68)/513 = 0.71

The actual building load on a 24 hour basis is 513 MJ.

Solar Fraction = (Energy provided by the solar system)/(total cooling load)

q_R = chiller output = (q_g x C.O.P.)

TABLE 7.11   Distribution of the Energy Supplied by the Solar Collector Array

Collector Area = 47 m^2

Items	Hours: Where the Collector Efficiencies > 10%										Total
	0800	0900	1000	1100	1200	1300	1400	1400	1600	1700	MJ/Day
q_ℓ (1) (MJ/hr)	16.6	16.6	16.7	18.3	19.9	25.0	30.2	30.8	31.5	36.3	q_ℓ (24 hrs) (513)
q_c (2) (MJ/hr)	5.5	36.7	58.4	74.2	89.3	85.3	74	62.3	43.5	19.6	(549)
Chiller (3) Rating q_R (MJ/hr)	2.5	18.4	35	35	35	35	35	35	29.6	11.6	(272)
$(q_R - q_\ell)$ (4)	14.1	1.8	18.3	16.7	15.1	10	4.8	4.2	1.9	24.7	
Hot Storage q_{h-s} (MJ/hr) (2) - (51.5 MJ/hr)	–	–	6.9	22.7	37.8	33.8	22.5	10.8	–	–	(135)
Cold Storage q_{c-s} (MJ/hr) (4) - (1)	–	1.8	18.3	16.7	15.1	10	4.8	4.2	–	–	(71)

Summary Table 7.11   Energy Flow Diagram

Collector daily output (q_c)   = 549 MJ/Day

Energy to hot storage (q_{h-s})   = 135 MJ/Day

Energy to cold storage (q_{c-s})   = 71 MJ/Day

Total energy through the chiller per day
           (q_R) = (272 + (135 x 0.68))   = 364 MJ/Day

q_c = q_g + q_{h-s}

Solar Fraction   =   0.71

Figure 7.10 shows the energy diversification and the collector inlet and outlet water temperatures for the 24 hour period.

CLOSURE

In the analysis of the absorption system, steady state operation was assumed. However, in real systems unsteady state conditions will occur, particularly at start-up and at periods where step changes occur either at the chiller input or output. Once unsteady state operation occurs, it can take over two hours for the system to return to steady state operation, so some form of "buffering" is usually provided to prevent the system from cycling. This can be achieved by incorporating both hot and cold storage in the system. Ideally the chiller should be undersized and always operated at full load when the chiller output exceeds the demand, the excess can be diverted to cold storage.

The equipment selected in the example is commercially available. The recommended collector area for the residential building of 47 m^2 to power a chiller of 7 kW capacity is compatible to that recommended by the solar air conditioning industry, that is, a collector area of 15-25 m^2 per ton of refrigeration (3.5) kW and storage of 30-60 litres per m^2 of collector area.

At present it appears that sufficient data are not available relating to the long term performance and economics of either residential or commercial solar powered absorption air conditioning systems operating in the Sydney region. A number of experimental systems have been operated in the region but data are inconclusive. Currently a demonstration system is in operation at the NSW Solar Energy Information Centre in Sydney [8] that uses 23 m^2 of evacuated tube collectors to power a 7 kW (2 ton) Yazaki chiller. The collectors are coupled to the hot water storage tank and produce hot water at temperatures up to 90°C for the chiller. A number of fan-coil units serviced by the chiller are used to air condition the Centre. The system can be programmed to switch from cooling to heating.

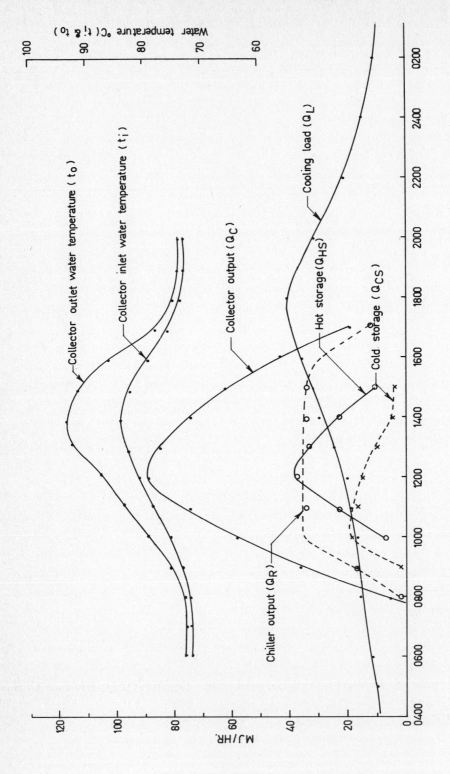

Fig. 7.10. Energy diversification and collector inlet and outlet water temperature vs. time for the residential building. Collector area 47 m² - chilled maximum capacity 35 MJ/hr.

Probably the best example of a solar powered Li-Br air-conditioning system is one that was recently commissioned by the Department of Housing and Construction for an office block building in Townsville [9].   This system consists of 203 m^2 of flat plate collectors coupled to two-35kW absorption chillers; 21 m^3 of hot water storage and 14 m^3 of cold water storage.   Auxiliary heat for the system is provided by an electric boiler.   A conventional vapour compression system can be used in series with the absorption chiller which is more efficient than operating the electric boiler at periods of low solar insolation.   A report on the system concludes that a hybrid system using a combination of solar powered refrigeration and vapour compression refrigeration could be energy efficient in the future.

NOMENCLATURE

Symbol	Meaning
A	is the heat transfer surface area (m^2)
A_c	the total collector area (m^2)
c_c	the specific heat of collector fluid (kJ/kg $^{\circ}C$)
C.O.P.	chiller coefficient of performance (Table 7.1)
f	solar fraction
G_T	the instantaneous solar radiation incident on the collector (W/m^2)
H_L	heat lost from the system (percent)
L	hourly cooling load (MJ/hr)
m_c	flow rate of collector fluid.
q	heat transferred (W)
q	Energy to match load (W)
q_a	heat loss (W)
q_c	energy output from the collectors (kW)
q_D	daily cooling load (MJ/day)
q_g	energy output from the generator (W)
q_s	heat transferred to space (W)
q_s	sensible heat load (W)
q_t	heat transmitted through glass (W)

q_{c-s} energy to cold storage (W)

q_{h-s} energy to hot storage (W)

q_L cooling load (kJ)

q_R energy to the chiller (W)

Q_{oa} volume flow rate of outside air (litres/sec)

T_a the ambient temperature of the atmosphere (^oC)

T_i collector inlet water temperature, ^oC

T_o outdoor design air temperature (^oC)

T_p the average temperature of the absorber plate (^oC).

T_s Water temperature at start of heat collection (^oC)

T' Water temperature at end of heat collection with no heat removed from storage (^oC)

ΔT_e equivalent temperature difference (^oC)

U overall heat transfer coefficient (W/m^2 ^oC)

W_{oa} moisture content of outside air (g/kg)

W_{RM} moisture content of return air (g/kg)

     the solar absorptivity of the glass (dimensionless).

ϵ emissivity of a surface

η_C instantaneous collector efficiency

η_D collector daily efficiency

     reflectivity of the surface

τ_{su} the solar transmissivity of the cover, i.e. the fraction of the solar radiation that reaches the absorber, (dimensionless).

REFERENCES

1. ASHRAE Handbook of Fundamentals, Published by the American Society of Heating Refrigeration and Air Conditioning Engineers, Inc., New York (1977).

2. Australian Department of Housing and Construction, Air Conditioning Systems Design Manual, Chapter 1-3, Australian Government Publishing Service, Canberra (1972).

3. "Residential HVAC Loads", a CP/M 2.2 computer program by ENGSOFT Pty. Ltd., Australia.

4. "Computerised Instrumented Residential Audit (CIRA)", a computer program for microcomputers, Publication 425R from Lawrence Berkeley Laboratories, Berkeley, CA (USA).

5. Ward, D.S. Vesaki, T. and Lof, G.O.G., "Cooling Subsystem Design in CSU Solar House III", Proc. Int. Solar Energy Soc. Conf., Winnipeg, vol. 3 (1976).

6. "Performance Data Catalog for Model WF 36 Chiller", Arkla Solaire Industries, Inc., Evansville, IN. U.S.A.

7. McCaffrey, J.J.,"A Review of Solar Powered Absorption Chiller Systems and Their Applications", M. Eng. Sci. Thesis, UNSW (1979).

8. Solar Energy Information Centre, Argyle Arts Centre, Sydney.

9. Chekalin, N. and Jones, P., "Solar Air-Conditioned Office in Townsville", vol. 4, No. 3, Solar Progress (1983) pp. 25-27.

CHAPTER 8

Sizing Space Heating and Cooling Components

INTRODUCTION

In this chapter the major components that constitute a space heating or cooling system are chosen. The component manufacturer's data and the system loads are an essential requirement for initial component sizing. The optimum position for auxiliary heating elements and temperature stratification in storage tanks are also discussed.

In sizing the components of the solar heating systems (such as those shown in Figures 8.1, 8.2 and 8.3), the collector area is the primary quantity to be calculated as it determines the sizes of most of the other components. The collector area may be chosen arbitrarily to provide a desired fraction of the total heat load or be optimised from an economic analysis to maximise the energy supplied per unit cost. Based on many computer based designs, experiments and experiences with existing installations, several rules of thumb have resulted that can be used as general guidelines in sizing solar system components for residential buildings. These "rules of thumb" were summarized in Table 6.2.

SELECTION OF A HEAT EXCHANGER

Heat exchangers may be needed at different locations in a solar home-heating system:

1. A heat exchanger is usually needed between the heating system and the domestic hot water system (Figure 8.2). This heat exchanger separates the water in the heating system, which may

contain corrosion inhibitors, antifreeze, etc., from the potable hot-water supply.

2. Figure 8.2 also shows a heat exchanger between the water in the solar heating system and the hot air supplied to the house.

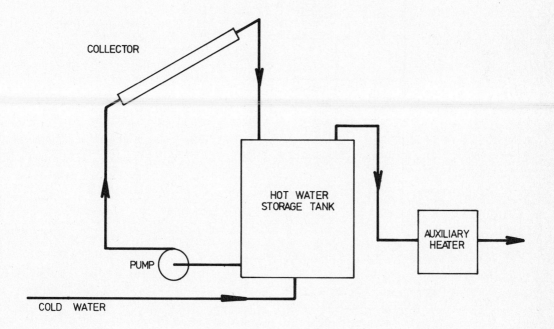

Fig. 8.1.   Basic liquid direct system.

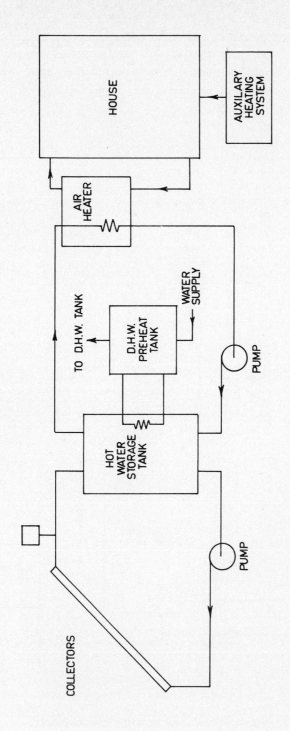

Fig. 8.2.   Diagram of a solar heating system.

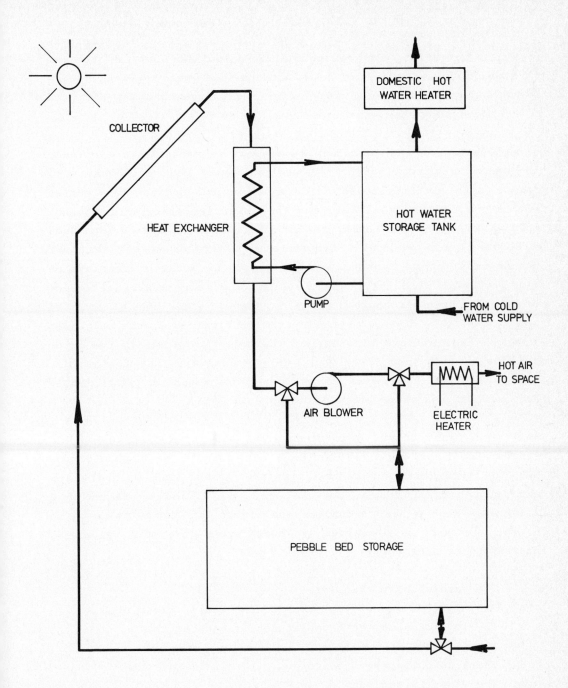

Fig. 8.3. Basic air system, domestic hot
water and space heating [4].

3. In regions where the water in the collectors may freeze, either antifreeze or a different liquid is used in the collector. Water is still the cheapest liquid to use in the thermal storage unit. A third heat exchanger is needed in this case to separate the two fluids. (This heat exchanger was not used in the system shown in Figure 8.2.)

To compensate for the increased losses, it is necessary to increase the area of the collector. A well designed heat exchanger will operate with a temperature difference of $5^{\circ}C$ between the outlet temperature of the cold fluid and the inlet temperature of the hot fluid. As a rule of thumb, the collector area should be increased by about 2% for every $1^{\circ}C$ temperature difference across the heat exchangers in items 2 and 3 above. Since the potable water system is in parallel with the heating system, the temperature difference across its heat exchanger (item 1 above) will not affect the collector size. Thus, for the system in Figure 8.2, the only heat exchanger that affects the size of the collector array is the water to air heat exchanger (item 2). In the analysis that follows, it is assumed that the water to air heat exchanger is sufficiently large so that the temperature difference is $5^{\circ}C$.

Shell and tube type heat exchangers are commonly used for transferring heat between liquids. The single pass counterflow type is recommended to reduce pressure losses and maintain low temperature differentials between the two fluids. Figure 8.4 illustrates such a heat exchanger and Figure 8.5 shows its location in the circuit. To select a heat exchanger, the following information is necessary:

a. The rate of heat transfer.
b. The type and quantity of each of the fluids circulated.
c. The inlet and outlet temperatures of each of the fluids circulated.
d. The allowable pressure drops.
e. The size limitations.

The surface area (A) in a heat exchanger needed to transfer the required heat is given by:

$$A = q/[U (LMTD)] \qquad\qquad (8.1)$$

where U is the overall heat transfer coefficient (usually between 0.85 and 1.7 kW/m^2 $^\circ$C for water to water heat transfer). The log mean temperature difference (LMTD) is given by

$$LMTD = [(T_1 - t_2) - (T_2 - t_1)]/\log_e[(T_1 - t_2)/(T_2 - t_1)] \qquad (8.2)$$

(5°C < LMTD < 8°C for well designed heat exchangers)

where the temperatures are defined in Figure 8.4.

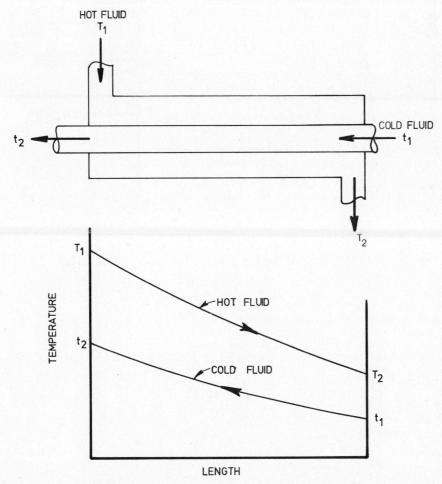

Fig. 8.4   Temperature variation along the the length of a single-pass counterflow heat exchanger.

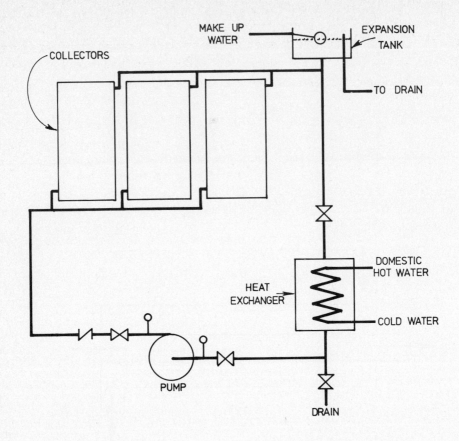

Fig. 8.5 Collector fluid loop with heat exchanger.

EXAMPLE 1 : Heat Exchanger Selection

Select a suitable heat exchanger for the collector loop shown in Figure 8.5.   The collector area is 30 m^2, the water leaves the collector at 70°C, the storage tank temperature is 55°C and the heat is transferred at the rate of 8 kW.

Solution:

To select a heat exchanger from Table 8.1 [6], calculate the heat transfer area (A) from equation (8.1).

TABLE 8.1   Shell and Tube Heat Exchangers, Tube Surface Area (m^2)   [6]

Series	Type HF-SSF One Pass Y Tubes	Type F-HF One, Two & Four Pass Y Tubes	R Tubes	Type SSF One & Two Pass Y Tubes	R Tubes	Four Pass Y Tubes	R Tubes
201	0.14	-	-	-	-	-	-
202	0.27	-	-	-	-	-	-
301	-	0.33	0.24	0.36	0.26	0.33	-
302	-	0.68	0.48	0.73	0.54	0.67	-
303	-	1.03	0.73	1.11	0.81	1.01	
502	-	1.68	1.04	1.71	1.10	1.51	1.07
503	-	2.52	1.56	2.58	1.65	2.32	1.62
504	-	3.36	2.08	3.47	2.22	3.10	2.17
602	-	2.5	1.63	2.54	1.70	2.38	1.54
603	-	3.74	2.44	3.82	2.57	3.58	2.32
604	-	4.98	3.24	5.11	3.43	4.78	3.10
606	-	7.51	4.90	7.73	5.19	7.24	4.70
608	-	10.1	6.53	10.33	6.92	9.68	6.27
802	-	-	3.17	-	3.04	-	3.04
803	-	-	4.70	-	4.57	-	4.57
804	-	-	6.24	-	6.11	-	6.10
805	-	-	7.78	-	7.64	-	7.64
806	-	-	9.30	-	9.16	-	9.16
807	-	-	10.84	-	10.70	-	10.7
808	-	-	12.38	-	12.23	-	12.23
809	-	-	13.90	-	13.77	-	13.77
810	-	-	15.44	-	15.30	-	15.30

$$A = q/[U \ (LMTD)]$$

The recommended values for U and LMTD are 1.28 kW/($m^2 \ ^\circ C$) and 6.5°C, respectively.

$$A = 8/[(1.28)(6.5)] = 0.96 \ m^2$$

From Table 8.1, choose the Series 303, Type F-HF, single-pass heat exchanger with Y-tubes with a tube surface area of 1.03 m^2.

Equation (8.1) gives the actual (or expected) LMTD for this heat exchanger

$$1.03 = 8/[(1.28) \ (LMTD)]$$
$$LMTD = 6.07^\circ C$$

EXAMPLE 2 :   Fluid Temperatures

Calculate the expected leaving temperatures for both fluids for the heat exchanger in Example 1.   The specific heat of the collector coolant is 3.34 kJ/kg $^\circ$C and its density is 1000 kg/m^3.

Solution:

For the collector fluid,

$$q_c = m_c c_c \ (T_1 - T_2) \tag{8.3}$$

where

m_c = flow rate of collector fluid. From Table 6.2 the recommended value is 0.014 litres/s per m^2 of collector area

and

$$m_c = (0.014) \ (30) = 0.42 \ litres/s \quad (0.42 \ kg/s)$$

In example 1, q_c was given as 8 kW, so

$$8 = (0.42)(3.34)(70 - T_2)$$

and

$$T_2 = 64.3^{\circ}C$$

for t_2, equation (8.2) gives,

$$LMTD = [(T_1 - t_2) - (T_2 - t_1)]/\log_e[(T_1 - t_2)/(T_2 - t_1)]$$

$$6.07 = [(70 - t_2) - (64.3 - 55)]/\log_e[(70 - t_2)/(64.3 - 55)]$$
$$t_2 = 60.5^{\circ}C$$

EXPANSION TANK SIZING

In a closed system where a liquid is subjected to varying temperatures, an expansion tank is necessary to prevent ruptures in the pipes and equipment. Expansion tanks can either be open or closed. A closed tank can be located anywhere in the system, whereas an open one must be located above the rest of the system or the liquid will overflow. For either case the tank must be properly sized. An undersized tank in a closed system will rupture and in an open system the liquid will overflow. On the other hand, if oversized, it will tend to be unduly expensive. To determine the proper volume of an expansion tank (V_t) in a water system the following equation can be used.

$$V_t = (V_w)(k)(T_h - T_1)(FS)$$

where

k = volumetric coefficient of expansion of water
(0.00018 m^3/m^3 $^{\circ}C$ between 5° and $100^{\circ}C$)
T_1 = lowest temperature attained by the water, $^{\circ}C$
T_h = highest temperature attained by the water
FS = safety factor to prevent overflow (usually
between 2 and 2.5)

The total volume of water (V_w) must include the water contained in all equipment and interconnecting pipes in the closed loop.

AUXILIARY HEATING

In heating or cooling a space the type of equipment used must be
carefully selected. If, for example, space heating using solar
heated water is considered, then an efficient method of transferring
the heat from the hot water to the space must be selected.
Conventional cooling coils used in central heating systems are
designed to operate at water temperatures in the range 65° to 75^OC.
To produce such temperatures using conventional flat plate
collectors would be inefficient and may even be impossible during
the winter months.   To provide the needed heat at lower water
temperatures would require larger surface areas (larger coil face
area) which may not be economical and may be difficult to
accommodate.   There are a number of solutions to this water
temperature limitation. One common approach is to use solar energy
to preheat the water and then to use an auxiliary heater to boost
the water temperature to between 60° and 65^OC.   If booster heating
is used, it may be applied at a number of positions.   Figure 8.6
shows a typical hot water space heating system. Applying the
auxiliary heat at position 1 is probably the most convenient and has
the lowest first costs.   However, since a larger mass of water is
heated than may be needed in a day this approach may impede solar
energy collection the following day.

If booster heating is used at position 2 in Figure 8.6., the space
heating system can be either all solar powered or it can be a
conventional space heating system.   The capacity of the heating
elements should be capable of supplying the total space heating
requirements if needed.

Locating the auxiliary heater at position 3 is probably the best
arrangement.   It has the advantages of 2 but it can also use the
solar collection system as a preheater, thus reducing the energy
input to the booster heater.

Precautions must be taken to prevent the auxiliary heaters from

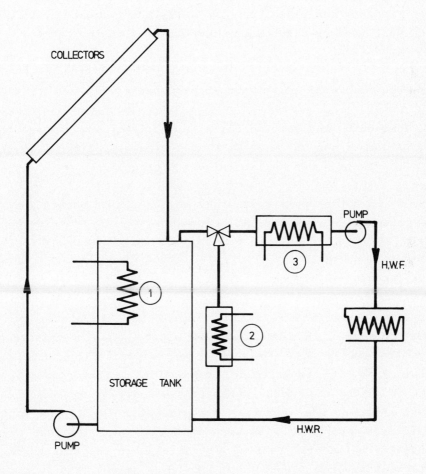

Fig. 8.6. Alternative positions for boost heating.

pressurising the system. High as well as low temperature cut-outs plus a pressure relief valve that is actuated by both temperature and pressure are necessary for any auxiliary heating source.

Other approaches that have been used to achieve a more effective space heating system are:

1.   Redesign of the coil, by increasing the fin size and reducing the pitch, to give satisfactory heat output per unit face area for water temperatures in the range 40° - 50°C.

2.   Use of oversized skirting-board heaters.

3.   Embedding the heating coils in the floor and heating the floor to say 25° - 28°C. The thermal capacity of the floor would assist in heat storage.

4.   Using a heat pump system whereby the solar heated water is the heat source for the heat pump.

TEMPERATURE STRATIFICATION

Temperature stratification will improve the performace of solar cooling systems. In both the hot water and the chilled water storage tanks it is desirable to maintain the differential temperature as high as practically possible. This requirement for example in hot storage tanks necessitates a height to width ratio of at least three to one. Both the position of the return water and the discharge rate should be such that circulation within the tank is minimal. Horizontal baffles may be used in the tank to minimise circulation in the tank. Figure 8.7 shows the temperature stratification that was achieved within one well-designed hot-water tank.

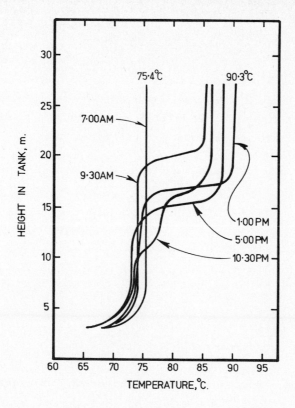

Fig. 8.7   Temperature profile in a storage tank as a function of the
time of day.

HEATING AND COOLING COIL SELECTION

Heating and cooling coils are usually selected from data for
standard coils provided by the manufacturers.   Figure 8.8 is
typical of the coil-rating data provided by manufacturers.   These
data are supplied by the manufacturer to help select a coil that
will meet the maximum heating load requirements.

EXAMPLE 1 -   Cooling Coil Selection

Using Figure 8.8, select a coil with a heating capacity (q) of 12.73
kW if the inlet water temperature (T_1) is 65°C and the inlet air dry

bulb temperature (t_1) is 20°C.   Assume that the water temperature decreases by 10°C as it passes through the coil.

Solution

To use the chart the hot water flow rate and the heat load must be determined.   An energy balance for the water gives:

$$q = \rho Q c_c (T_1 - T_2) \qquad\qquad (8.3)$$

where ρ is the density of the fluid, Q is the volumetric flow rate (in litres per minute), and c_c is the specific heat of water (4.19 kJ/kg$^{\circ}$C)

so,

12.73 = 1000 Q (4.19) (10)/[60(1000)]
Q = 18.23 litres/min   (3.04 x 10^{-4} m^3/sec)

From Figure 8.8, for Q = 18.23 litres/min, curvers A to D are intersected at a Heat Transfer Index (H.T.I.) of approxiately 181, 269, 340 and 460 respectively.   The H.T.I. is a function of the heating capacity (q) of the coil, the entering hot water temperature (T_1), and the entering air dry bulb temperature (t_1):-

where
$$q = (H.T.I.) (T_1 - t_1) \qquad\qquad (8.4)$$

If T_1= 65°C and t_1 = 20°C, then the heating capacity of each coil is:-

q(Coil A) = 181 (65 - 20) = 8.145 kW
q(Coil B) = 269 (65 - 20) = 12.105 kW
q(Coil C) = 340 (65 - 20) = 15.300 kW
q(Coil D) = 460 (65 - 20) = 20.700 kW

Since the heating load is 12.73 kW, Coil C is the best choice.

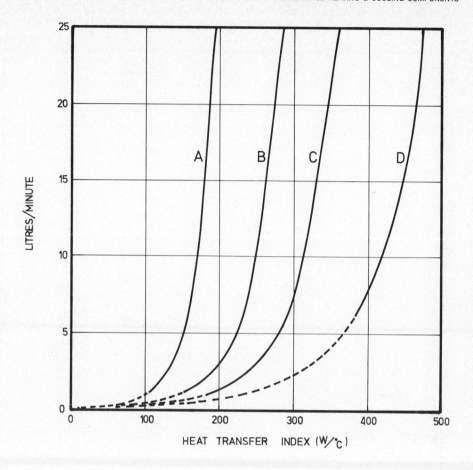

Fig. 8.8   Coil hot water rating.

PRESSURE DROP CALCULATIONS

Once a coil is selected, the next step is to check the pressure drop
for the water across the coil.   Figure 8.9 shows the pressure drop
rating of the above coils.   For coil "C" with a water flow rate of
18.23 litres/minute, the figure indicates a pressure drop of 42 kPa.
This flow resistance must be added to the resistances of the circuit
pipes and fittings.

If the pump head (H) and the efficiencies of the pump (η_p) and
motor (η_m) are known, the input power (P) to the pump can be
determined from:

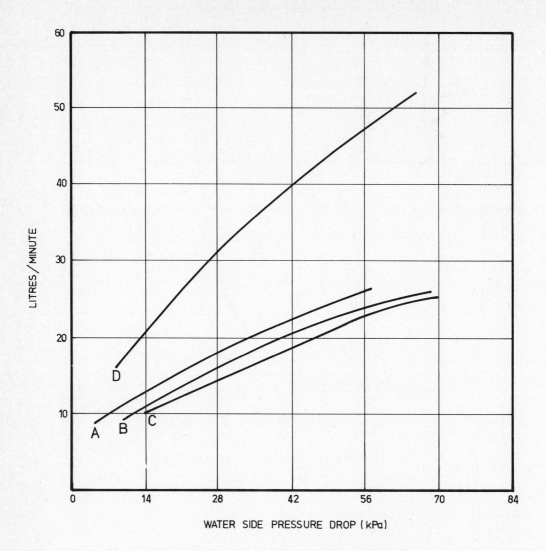

Fig. 8.9 Coil pressure drop rating.

$$P = \rho \, Q \, H / [\, \eta_p \quad \eta_m] \qquad\qquad (8.5)$$

Considering the coil head loss only, for a flow rate of 18.27 litres/minute and a pressure loss of 42 kPa, and the efficiencies of the pump and motor are 0.6 and 0.8, respectively,

$$P = 0.000304 \ (m^3/s) \ x \ 42 \ (kN/m^2)/(0.6 \ x \ 0.8)$$

$$P = 26.6 \text{ Watts (Electric power required to circulate}$$
$$\text{the hot water through the coil)}.$$

AIR SIDE CONSIDERATIONS

The previous calculations relate only to the "water side" of the coil. In addition, there must be sufficient air flow through the coil to transfer the 12.73 kW of heat from the coil to the space to be heated. The air flow rate can be determined by an energy balance. For a heating load of 12.73 kW the mass flow rate of air (m_a) may be calculated from:-

$$m_a = q \ /[c_p \ (t_2 - t_1)] \qquad\qquad (8.6)$$

where c_p is the specific heat of air (1.00 kJ/kg^{o}C) and t_1 and t_2 are the air inlet and outlet temperature. If air enters the coils at 20° and leaves at 30°C, then

$$m_a = 12.73 \text{ kW}/[1.00 \ (kJ/kg^{o}C) \ x \ (30 - 20)^{o}C]$$
$$m_a = 1.273 \text{ kg/s}$$

Thus, for these conditions the required air flow is 1061 litres/second for air at a density of 1.2 kg/m^3.

SUPPLY AIR DUCTS

Recommended air duct velocities should be between 3 and 6 m/s. For a flow of 1016 litres per second (as in the example above) this velocity range will require duct areas of between 0.177 and 0.354 m^2, i.e., duct sizes of 420 x 420 mm and 595 x 595 mm.

It is usually not economic for residential fan-coil units to have separate hot and cold water coils. In this example the same coil will be used for both heating and cooling. The performances of the coils used in the previous examples in the cooling mode of operation are shown in Figure 8.10 [5]. This figure is based on the following conditions:

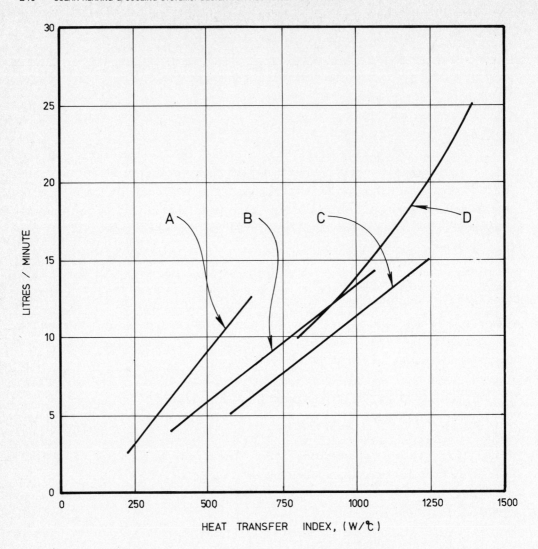

Fig. 8.10 Coil chilled water rating.

Inlet chilled water temperature $T_i = 7^{\circ}C$
Apparatus dew point temperature $T_a = 12^{\circ}C$

As an example, consider a coil which has a cooling load of 6 kW.
The Heat Transfer Index from equation (8.4) is:

H.T.I. = 6000 W/(12 - 7)$^{\circ}$C = 1200 W/$^{\circ}$C

For coil "D" if the H.T.I. is 1200 W/$^{\circ}$C and the chilled water flow
rate is 17.5 litres/minute the exit water temperature will be 12°C.
In this case, the pressure drop is 9.8 kPa (Figure 8.9) and the
power required would be 6 Watts (coil pressure drop only
considered.)

Air Side Requirements

An air flow rate of 1061 litres/second was used to meet the heating
load. For the space cooling requirements it is necessary to use a
psychrometric chart to calculate the air flow rates. Indoor and
outdoor design conditions plus the amount of outside air flowing
through the conditioner must be known or estimated. An estimate of
the coil by-pass factor for the operating conditions is also
required. The coil bypass factor is a measure of the amount of air
that passes through the coil without any change in temperature or
relative humidity. It is a function of the coil design and the
velocity of the air through the coil. A detailed analysis of
cooling coil performance and sizing which includes coil bypass
factors is given in reference [3] and [7].

SOLAR AIDED HEAT PUMP

A heat pump is a device used to transfer heat from a low temperature source to a high temperature sink.   If the purpose of the device is to cool, it is commonly called a "refrigerator".   On the other hand, if it is used for heating, the device is called a "heat pump". A heat pump can provide either cooling or heating in the same region by a simple reversal of the flow direction of the refrigerant, hence these devices are also called "reverse cycle" refrigeration systems.

To operate effectively as a "heat pump", a dependable source of heat must be available at reasonably high temperatures.   The coefficient of performance (C.O.P.), which is a measure of the effectiveness of a heat pump, is defined as the ratio of the energy utilised to the energy paid for.

The C.O.P. for an ideal Carnot cycle can be used to estimate the C.O.P. of real cycles.   The Carnot C.O.P. is given by,

$$C.O.P. \ = \ T_H/[T_H - T_L]$$

where T_L refers to the temperature of the heat source and T_H to the temperature of the heat sink.   This relationship shows that an increase in T_L will result in an increase in the C.O.P. of the system.   Since the heat source could be solar energy, a solar aided heat pump is potentially a viable system.

NOMENCLATURE

Symbol	Meaning
A	is the heat transfer surface area (m^2)
A_c	the total collector area (m^2)
c_c	the specific heat of collector fluid (kJ/kg ^oC)
c_p	the specific heat at constant pressure of air (kJ/kg ^oC)
C.O.P.	chiller coefficient of performance (Table 7.1)
FS	the safety factor to prevent overflow
G_T	the instantaneous solar radiation incident on the collector (W/m^2)
H	head loss (kPa)
H.T.I.	heat transfer index ($W/^oC$)
k	volumetric coefficient of expansion of water ($^oC^{-1}$)
LMTD	the log mean temperature difference (^oC)
m_a	mass flow rate of air (kg/s)
m_c	mass flow rate of collector fluid (kg/s)
P	power (W)
Q	volumetric flow rate (litres/sec)
T_a	the ambient temperature of the atmosphere (^oC)
U	overall heat transfer coefficient (W/m^2 ^oC)
V_t	volume of the expansion tank (m^3)
V_w	the total volume of the water (m^3)
η_m	efficiency of motor
η_p	efficiency of pump
ρ	fluid density (kg/m^3)

REFERENCES

1. Colorado State University Solar Energy Applications Laboratory, "Solar Heating and Cooling of Residential Buildings. Sizing, Installation and Operation of Systems", U.S. Government Printing Office (1977).

2. Colorado State University Solar Energy Applications Laboratory, "Solar Heating and Cooling of Residential Buildings, Design of Systems", U.S. Government Printing Office (1977).

3. Carrier International Ltd., "Handbook of Air Conditioning Design", McGraw-Hill, New York, (1965).

4. Yvonne Howell and Justin Bereny, Engineers Guide to Solar Energy, Solar Energy Information Services, San Mateo, California, 94401, (1979).

5. Carrier Catalog, "Fan-Coil Weathermakers Model 36 QA", No. 36QA/100-10/64, KRALCO Printers, Sydney.

6. Young Radiator Co., "Fixed Tube Bundle Heat Exchangers", Catalog No. 1275, Wisconsin (1975).

7. Australian Department of Housing and Construction, "Air Conditioning System Design Manual", Chapters 1-3, Aust. Govt. Pub. Serv., Canberra (1972).

CHAPTER 9

Ancillary Equipment

INTRODUCTION

In this chapter, the important 'standard' components of solar energy systems will be considered. Such components as fans, pumps, valves and controls have much in common with those used for heating and cooling from conventional energy sources. However, there are significant differences. These differences can determine whether a solar energy system is effective or functionally a failure.

Firstly, the circulating systems and their primary components, pumps and fans, will be described. Next, the control devices required to ensure effective system operation will be discussed. Temperature extremes which a collector system can encounter may lead to, on the one hand, a tendency for water contained in collector channels and pipes to freeze, on the other hand if water stagnates in certin collectors overheating may occur. Any solar energy system design procedure should include a check for these possibilities and should result in adequate safeguards against their occurrence.

There are very few differences between solar heating practice and conventional fuel heating practice with regard to reticulation of heated air and water.

All air and water reticulation systems need to be designed having due regard to energy losses and other undesirable features such as noise. Reference will be made to recommended procedures for pipe sizing and system layout.

The system designer should not attempt to take the short-cut of guesstimating overall system losses. Not only are vital components, such as collector, pump or fan, likely to be wrongly sized. The very discipline of working through the system resistance, component by component, identifies regions of high loss and might well lead to

system   simplification   or   component   re-design   which   would
considerably improve system performance.

Useful sources of information for sizing of pump installations are
provided in references [1] and [2].   For fans an even more
comprehensive set of data is given in reference [3] although
unfortunately imperial units are used in this case.

PUMPS

Active (i.e. non thermosyphon) liquid solar energy systems require a
circulating pump. An initial distinction here must be made between
open systems, in which the usable water is circulated directly
through the heater, and closed systems which recirculate distilled
water, an antifreeze or a heat transfer fluid.   In locations where
water is particularly corrosive or where freezing temperatures
occur, closed systems are usually indicated.

The material and construction of the pump, i.e. its impeller,
casing, motor, bearings, shaft and seals will all depend somewhat
upon the nature of the circulating fluid.

Centrifugal pumps are best suited for almost all solar energy
applications from swimming pools to space heating. Direct coupled
motors are most suitable and if these can be of variable speed the
control and efficiency of the system can more readily be optimised.
Centrifugal pumps, if correctly installed, present few mounting or
balancing problems.   Noise and vibrations are minimal, efficiency
is generally acceptable and reliability is generally good.   Should
a discharge line be blocked a well-designed centrifugal pump will
continue to operate against an increased head and create little
likelihood of system damage or overpressure.

A typical centrifugal pump characteristic curve is shown in Figure
9.1 for a specific motor speed. Although centrifugal pumps have a
degree of tolerance to system mismatch or incorrect flow selection,
it is most desirable for an accurate estimate of flow and system
characteristics to be made.

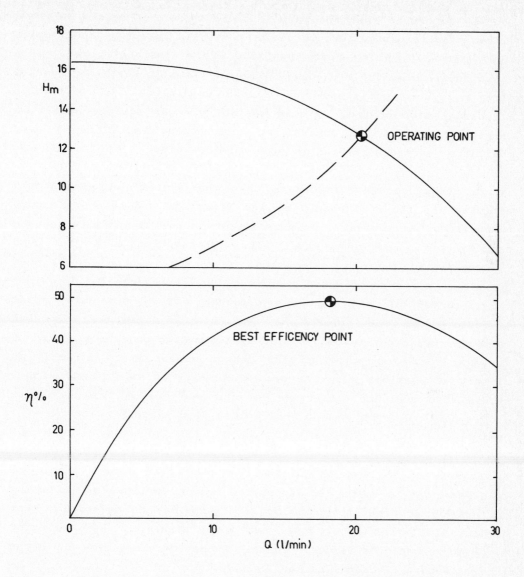

Fig. 9.1.   Typical centrifugal pump characteristic.

It is often considered that the usual prudent engineering
application of generous safety factors should also be applied to
pump flow selection.   In fact, such an approach is potentially a
recipe for disaster.   Because of the way that the pump
characteristic falls off, if the pump is underrated on flow its head
delivery and efficiency will deteriorate rapidly.   However, it is

equally important that the pump should not be over-rated.   If too little flow is passed through the pump it will operate in the region near shut-off which can manifest flow instabilities (hence vibrations, noise and mechanical damage).   It will certainly operate with a reduced efficiency under such conditions.   If excessive discharge valve control is used instability can again result;   if excessive inlet valve control or motor speed variation is used cavitation may result which is even more damaging.

It is therefore most important that the system designer should estimate the required head rise and flow rate and select the pump accordingly.   A typical system resistance curve is indicated by the dashed line in Figure 9.1.   Its intersection with the pump characteristic curve is the system operating point.   The system resistance curve is obtained by summing the resistance (head against flow) curves of each system component.   Such data for solar collectors and for filters may be obtained from the component manufacturers or by testing.   The system design objective should be for the operating point at the design condition to coincide as closely as possible with the best efficiency point.   The designer should also evaluate resistance for the extremes of flow to ensure stable pump operation in the anticipated flow range.

A variable speed pump or at least a two speed device should be considered as a means of improving system efficiency.   During periods of low insolation a low flow rate may be selected.   For higher insolation the flow rate is raised;   the collector temperature would be reduced because of the increased flow and hence the collector efficiency raised.

Pumps are specified by the flow rate (capacity) they will provide when subjected to a given increase in head.   In a closed loop, the head or pressure is the sum of the pressure drops (friction losses) in each component in the loop which could include the connecting pipes, valves, collectors, heat exchangers, etc.   In an open system, the pressure also includes the pressure necessary to bring the liquid to the higher elevation.

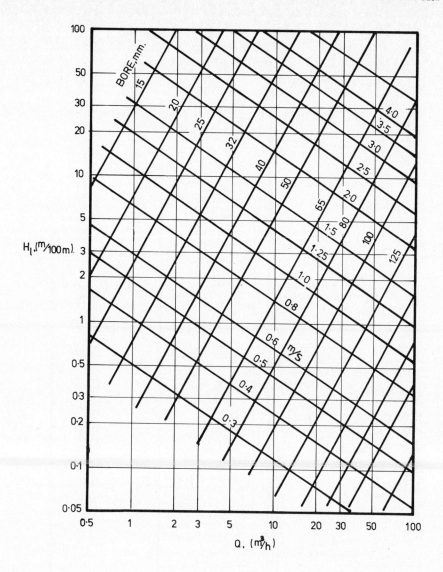

Fig. 9.2.   Loss of head for water flowing in straight pipes.

Figure 9.2 may be used for determining friction losses in iron or steel pipes and Figure 9.3 for approximating the length of a straight pipe that will have the same friction loss as a given fitting.   The following example demonstrates the use of Figures 9.2 and 9.3 for sizing a pump system.

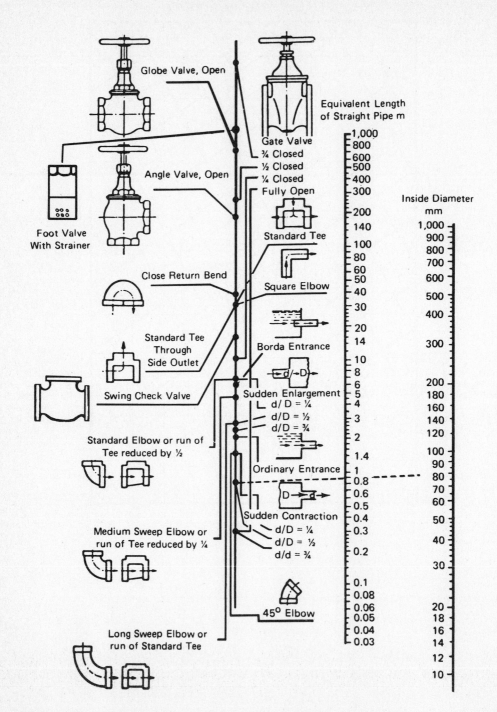

Fig. 9.3. Approximate head losses due to friction in pipe fittings
(courtesy APMA [2]).

TABLE 9.1   Recommended Water Velocity Based on Minimising
             Noise and Erosion while Maintaining Economical Design [4]

Service	Velocity Range m/s
Pump Discharge	2.5 - 3.7
Pump Suction	1.2 - 2.1
Drain Line	1.2 - 2.1
Header	1.2 - 4.6
Riser	0.9 - 3.1
General Service	1.5 - 3.1
City Water	0.9 - 2.1

EXAMPLE 1 :   Pipe and Pump Sizing

(a)   Determine the size of the main interconnecting pipes.
      From Figure 9.2 for the system in Figure 8.5:-
      flow rate = 0.42 litres/s = 1.512 m^3/hr
      assumed pipe fluid velocity < 1.5 m/sec (see Table 9.1)
   Choose 20 mm bore pipe with an expected pressure loss of
   16 mm of water per 100 m of equivalent length (Figure 9.2).

(b)   Determine the capacity and head of the centrifugal pump to
      use in the collector fluid loop.

Solution

(a)   The head developed is the head loss by all components in the
loop that are connected in series.

      Head developed = 16 x $\dfrac{\text{total equivalent length}}{100}$

For equivalent lengths of components and fittings (Figure 9.3)

Component or Fittings	Equivalent Length
3 gate valves (fully opened)	3 x 0.13 = 0.4
1 check valve	1 x 1.00 = 1.0
5 std. elbows	5 x 0.65 = 3.3
3 tee	3 x 1.4 = 4.2
1 strainer	1 x 8.0 = 8.0 (assumed)
1 heat exchanger	1 x 12.0 = 12.0 (assumed)
1 collector module	1 x 16.0 = 16.0 (assumed)
straight pipe	= 30.0 (approx.)
Total equivalent length	74.9

Head loss = 16 x 74.9/100   = 11.98 m of water

(b) Use centrifugal pump capacity = 0.42 litres/second

head   = 11.98 m of water

BLOWERS

Many of the fundamental considerations governing the selection of blowers or fans for air heating systems correspond with those for circulating pumps.   Once more it is essential to accurately estimate the required pressure rise and flow capacity. If there is an error in this rating of more than, say 20%, not only will the efficiency suffer drastically but flow instabilities could be set up which could easily damage the blower. This is especially true in the case of axial flow fans.

Compact packages which contain blower (often of the forward-swept centrifugal type), motor, dampers, instrumentation and controls as an integral unit are favoured in the heating, ventilation and air conditioning industry.   Such units save the costs and problems of excessive field installation.   Also, belt drives are more likely to be recommended for warm air blowers than direct motor coupling, since the motor would have a shorter service life in warm air.

Some of the solar space heating systems that use air-cooled collectors employ two blowers rather than one [5].   A 'collector' blower is used only when the solar heat is being collected, either for storage or for direct use. A 'load' blower supplies warm air to the rooms either from the collector or from storage. Both blowers operate when the building is being heated from the collector. Although the two blower system is more expensive than the use of a single blower it has an overriding advantage that the air flow rate through the rooms need not be limited to that desired for the collector and the control; also the ducting systems are simpler. Furthermore, most conventional heating units (the oil or gas fired unit used for stand-by or boost operation in conjunction with solar energy) already have a blower which will provide the load function.

CONTROLS

The three basic components of a control system are the input transducer or sensor, the controller and the output transducer. The function of the control system is to ensure that as much heat as possible is collected and delivered to the building on demand.   The controller can have a large effect on the overall efficiency of the system.

Most collectors for solar energy applications are presently on/off controllers.   The typical control strategy would be to start the pump or fan whenever the collector fluid exit temperature is greater, by some set temperature difference (ΔT_{ON}) than the tank

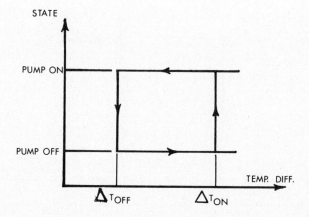

Fig. 9.4.   Representation of hysteresis
in the differential thermostat.

temperature.   The circulating device is turned off whenever the
collector outlet temperature falls to within ΔT_{OFF} of the tank
temperature.   This sequence is illustrated schematically in Figure
9.4.   For water systems it is usual to set ΔT_{ON} to $10^{O}C$ and ΔT_{OFF}
to $2^{O}C$.

As an example consider a hot water system in which the storage
temperature is $42^{O}C$ and the collector temperature is $20^{O}C$ when the
sun rises.   The collector temperature will gradually increase and
when it reaches $50^{O}C$ (the storage temperature meanwhile having
fallen to 40°) the controller will command the pump to start (see
Figure 9.5).   In the afternoon as the sun begins to set the
collector temperature will begin to decrease.   Suppose the storage
temperature has reached $60^{O}C$ by 4.00 p.m.   When the collector
temperature decreases to $62^{O}C$ the controller will stop the pump.

A simplified circuit for a single function differential thermostat
(controller) is shown in Figure 9.6.   The function of the feedback
loop is to regulate the hysteresis, represented schematically in
Figure 9.4, and to prevent system instability or cycling.

The temperature detected by a sensor depends on its location in the
system.   If the temperature sensor for the collectors is located at

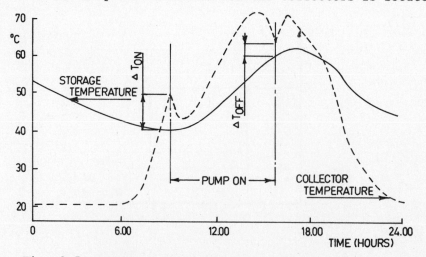

Fig. 9.5.   Control sequence for hot water systems.

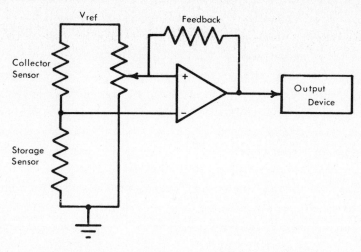

Fig. 9.6. Simplified control circuit.

a point where it is rapidly cooled by the fluid this can result in a
rapid on/off cycling of the pump. Such cycling should be minimised.
It causes wear on pumps, motors, electromechanical control
components and can also lead to noise disturbances. Cycling tends
to occur if the ratio between on and off temperature differences is
not properly selected.   This ratio should be between 5 and 7.   The
ratio in the example above was 5.   A larger value of the ratio will
reduce the total energy collected while a smaller value will carry a
risk of system instability and excessive cycling.

The thermostat should be the only control calling for manual
intervention by the building occupant.   For space heating
applications a two-stage heat indoor thermostat is recommended.   For
solar heating and cooling systems a two-stage heat, one stage cool
type is recommended.   The function of the second stage on heating
operation is for activation of the auxiliary heating system should
solar heating input be inadequate.

The type of sensor to be used should be selected carefully in
conjunction with the controller.   Most control systems use
thermistors.   The output ranges of these, typically 0-10 volts, are
sufficiently high for direct use in the controller.   Thermistors
give a non-linear output which must be compensated for in the

control circuitry. Liquid or vapour expansion elements are usually rejected as they have a relatively short operating life. Thermocouples are recommended for precision measurement and might lend themselves to the commissioning of an experimental installation. However their output signal is a low level one, in the millivolt range, and thus provision for amplification (and also for cold junction referencing) must be made. These thermocouple systems tend to be expensive.

Increasing use is being made of solid state ON-OFF controllers having a shielded thermocouple as a sensing element. These controllers can be set to switch to within $\pm 1/2^{\circ}C$.

Semiconductor devices now in use as temperature sensors offer promise of sensitive temperature control whilst lending themselves well to incorporation into simple integrated circuit controllers. A hybrid controller incorporating silicon devices, a thermistor and a relay has been described in reference [6]. The control of solar energy system operation presently allows scope for improvement by designers offering improved sensitivity and stability at lower cost.

Consideratons for solar pool heating controllers and sensors are different from those for domestic water and space heating or cooling. For pool heating the quantity of direct and diffuse radiation, the wind speed and the presence of rain all determine whether the water passing through the collector panels will be heated or cooled.

If a sensitive (to less than $1^{\circ}C$) temperature difference detector is available this can be used to detect the temperature difference between pool and supply line.

An alternative possibility is the use in the collector area of a thermostatic sensing device contained within a matt black casing. The casing is so configured that a composite 'sol-air' temperature is detected in such a way that the detector responds to weather in a similar manner to the collector panels.

SYSTEM PROTECTION

FREEZE PROTECTION

Although freeze protection of solar water heating equipment in the Sydney area may be considered unnecessary only one severe frost is required to bring the life of the equipment to an abrupt end. Furthermore, over much of the area of Australia and New Zealand freeze protection is mandatory.

Freezing of water in the collector could cause burst pipes and other mechanical damage. The collector inlet pipes are particularly vulnerable since, if these freeze, local boiling may still take place within the collector giving a near certainty of collector damage. There are three principal means of alleviating such problems:-

1. use of antifreeze or a special heat transfer fluid,
2. use of electric heaters or recirculation of warm water during period of frost,
3. draining of the system.

Anti-freeze Agents

Only closed circulation systems in which there is no possibility of contamination of consumable water by circulating fluid can be considered for this solution. The use of an auxiliary heat exchanger around the storage tank (see Figure 10.7) is indicated. The use of an auxiliary heat exchanger is also necessitated by economic considerations since provision of 100 litres or more of antifreeze would be prohibitively expensive   Many U.K. and U.S. systems use a special pre-heat cylinder for this purpose. An additional pump is usually required for such a system.

The working fluid should be selected in the context of an overall system evalution, including collector materials. The basic choice is between the addition of a conventional anti-freeze agent, such as ethylene glycol, to the circulating water and the use of special

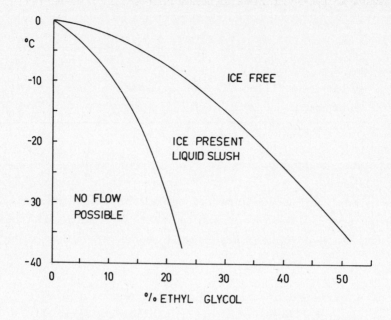

Fig. 9.7. Effect of various volumetric
concentrations of ethyl glycol.

heat transfer fluids which confer additional benefits upon the solar
heat collection process.

Automotive grade ethylene glycol solution is generally suitable as
an anti-freeze agent (see Figure 9.7). This usually contains a
corrosion inhibitor which is beneficial, especially in solutions
above 30% volumetric concentration. The chemical compatibility of
the corrosion inhibitor with the collector material should be
established, however, and especial care taken when the material is
aluminium. Thermal decomposition of ethylene glycol is potentially
a problem and can occur at temperatures as low as 120°C if air is
present in the system (e.g. through vent pipes).

Silicon oils, such as polydimethylsiloxane, should also be
considered when freezing is likely to be a problem. They have good
heat transfer characteristics, will not freeze and may be used in
all-aluminium collectors.

Re-circulation or Heating

Another approach to freeze protection is to heat the collector
sufficiently to avoid sub-zero temperatures [7].   Either the warm
water from the storage tank can be recirculated during the frost
period or additional electric resistance heating may be provided.
For protection against occasional or mild frosts the economic
penalties of the above methods are not severe.   In colder locations
the penalties may be excessive.   When the additional heating is
provided by recirculation of water from the storage tank the heat
gained during periods of collection is simply debited with this
required warm water.   If this method is adopted additional storage
capacity may also need to be provided.   When electric heating is
used the amount required is added to the boost requirement.
However, if a freeze were to occur in a normally mild region such as
Sydney the local demand for electricity would increase significantly
and it is likely that power outages would occur.   In that case,
these recirculation and electric heating techniques would probably
not work.

Draining of the System

In pumped systems the most direct method of frost protection is for
the collector to drain to the storage tank whenever the pump is
turned off.   A vent is provided at the top of the collector so that
air can enter the collector as the water drains out.   The collector
tubing must be arranged in such a way that all water drains out when
the pump stops and no stagnant pockets remain.   This simple system
is illustrated in Figure 9.8.   The dashed line represents the water
level when the pump is not operating.

A similar principle may be used with a thermosyphon collector.   To
avoid draining the storage tank, thermostatically actuated valves
must close the pipe between collector panel and storage tank when
freezing occurs.   A drain valve below the collector and an air vent
valve above it must open.   After the frost danger period has passed
the sequence is reversed.   Many people consider that the complexity
of such a system defeats the object of a thermosyphon system.
Furthermore, it is precisely during a heavy frost that control valve
operation is most vulnerable.

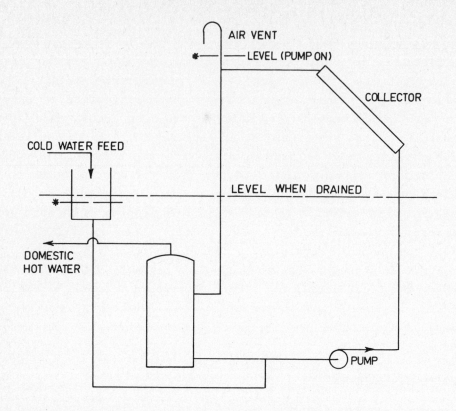

Fig. 9.8. Direct pumped system (self draining).

OVERHEAT PROTECTION

In addition to the normal differential control, protection against excessive water temperatures is necessary. If, for example, a power failure occurs during sunny weather, boiling could occur due to the loss of circulation. All systems should be designed with appropriate vents and relief valves to permit discharge of high pressure steam under such conditions. A problem then occurs in a closed system when power is restored. The thermal shock that would occur when the relatively cool water first reaches the overheated collector could damage the system. Also, there could be insufficient fluid remaining to prime the pump and regain circulation. For protection against these possibilities a high temperature limit switch and some provision for make-up fluid is recommended.

If the circulating fluid is air or a high-boiling point fluid a temperature limit switch could simply discontinue the pump operation.   The collector temperature would rise substantially; the collector would need to be constructed from suitable high-temperature materials.   In addition, it might still be necessary to provide venting of steam for the storage system.

The provision of constant temperature water for domestic use is best arranged by means of a thermostatically-controlled mixing valve. The desired temperature is obtained by admitting cold water to the hot water delivery downstream of any boost heating provision.   The solar hot water tank is allowed to reach any temperature attainable, collector temperature being only limited by the release of steam generated through a relief valve.   If the temperature of the service tank drops below the desired temperature boost heat is added.

A less efficient alternative is to set a temperature limit control on the solar storage tank, providing the maximum desired temperature of domestic hot water.   Because solar-heated water will not be delivered at a temperature higher than the set point less solar-heated water will be collected and stored.

REFERENCES

1. Klein, Schazlin and Becker, "Pump Handbook",
    Aktiengesellschaft, Frankenthal (1968).

2. "Australian Pump Technical Handbook", Australian Pump
    Manufacturers' Association Ltd., (1980).

3. Jorgensen, R. (ed.) "Fan Engineering", 7th Ed., Buffalo Forge
    Co., Buffalo, N.Y., (1970).

4. "Handbook of Air Conditioning Design", Carrier International
    Ltd., McGraw-Hill, New York (1965).

5. Lof, G.O.G., "Systems for Space Heating with Solar Energy"
    in "Applications of Solar Energy for Heating and Cooling of
    Buildings", Eds. R.C. Jordan and B.Y.H. Liu. ASHRAE GRP 170.

6. Czarnecki, J.T. and Read, W.R.W. "Advances in Solar Water
    Heating for Domestic Use in Australia", vol. 20, Solar Energy
    (1978) p.75.

7. Wilcox, B.A. and Barnaby, C.S. "Freeze Protection for
    Flat-Plate Collectors using Heating", vol. 19, Solar Energy
    (1977) p.745.

CHAPTER 10

Equipment Specification and Installation

INTRODUCTION

In this chapter the procedures of overall system planning, equipment specification, and installation are addressed. Different considerations apply to the integrated design of the solar collection system or to a retrofit situation. The quality of an installation and its lifetime performance are often largely dependent on the original specifications and the installation procedures adopted.

INTEGRATED DESIGN

It has been customary to classify buildings designed to use solar energy as either 'active' or 'passive' installations. Such a classification is not particularly helpful for two main reasons. Firstly, in considering space heating and cooling there are many homes and buildings which fall into a hybrid area between 'active' and 'passive'. Examples are the Trombe wall, shown in Figure 1.7, and the Hay roof pond in Figure 1.9. Secondly, the use of the adjective 'passive' in this context is only appropriate to cold and temperate climates and is not relevant to tropical or sub-tropical zones where the problem is one of providing a cooling rather than a heating effect.

For these reasons the objectives of architectural design for most effective use of solar energy would best be subsumed under the use of some such title as 'minimum energy housing'.

If the design of buildings is viewed in this way and the sums are performed correctly the economic evaluation of solar energy utilisation is transformed. This may also result in changing aesthetic perceptions of building design.

RETROFIT

Although some old buildings might justify a radical redesign, featuring a major solar energy installation, in general the approach to existing buildings will be one of retrofitting for partially fulfilling energy supply requirements. It is conceivable that some such applications could include space or process heating and cooling but the majority of such installations will be for domestic water heating.

Collector installation depends upon available roof, wall or ground space. If suitable roof area is not available the collectors will need rack or stand-off mounting. Rack mounting is suitable for flat roof, wall or ground application and consists essentially of a triangulated frame (designed to sustain dead loads and wind loads). Stand-off mounting is for a roof having the 'wrong' slope or orientation and consists of support members which allow air and rainwater to pass underneath, thus minimising problems from mildew, rust, rot and leakage.

Collectors can be mounted directly onto the roof surface. It is necessary to introduce a waterproof membrane between collector and roof and waterproof sealing beneath and around the collector.

Since installation and plumbing costs are high, especially between collector and storage tank, it is desirable to incorporate as much of the plumbing as possible in an integrated package. This is seen as a principal reason for the economic effectiveness of the integral collector and storage tank systems produced by several manufacturers in Australia.

A packaged unit is purchased at a price which has been kept low by standardised mass production. The unit is installed without the need for professional advice and with a minimum of on-site plumbing and electrical work. The objective should be a well-engineered standardised integral unit which will minimise on-site work.   In the final analysis, the life cycle costs for the system will depend primarily on the durability of the units.

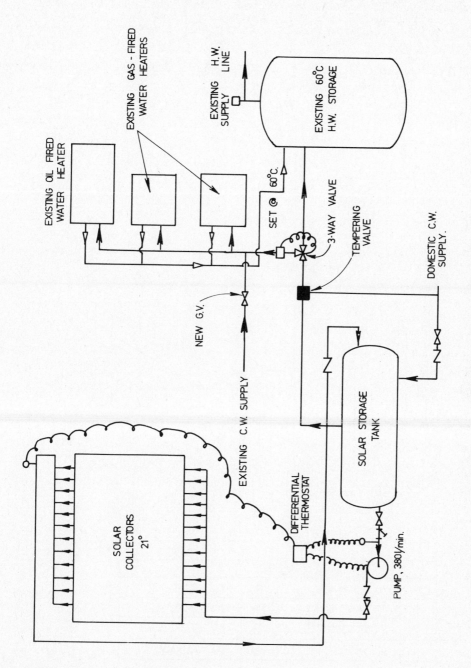

Fig. 10.1. Piping and equipment diagram for basic indirect system.

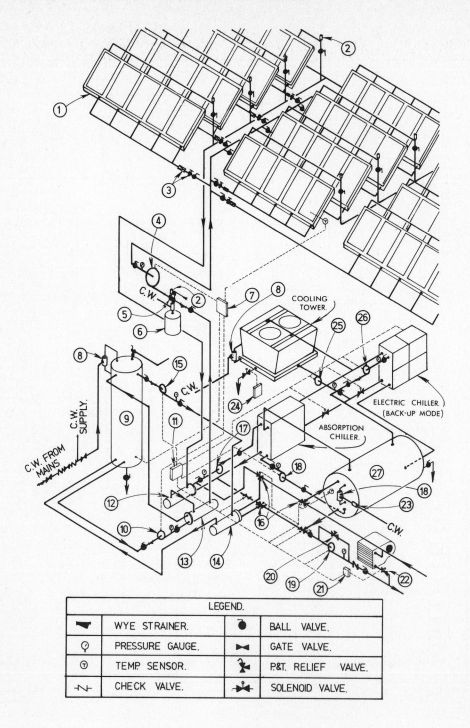

Fig. 10.2. Solar assisted heating and cooling system.

	LEGEND.			
⊷	WYE STRAINER.	●	BALL VALVE.	
℗	PRESSURE GAUGE.	⋈	GATE VALVE.	
Ⓣ	TEMP. SENSOR.	⚡	P.&T. RELIEF VALVE.	
⊣⊢	CHECK VALVE.	⊠	SOLENOID VALVE.	

WORKING DRAWINGS

Pipe working drawings together with written specifications are
needed to:

  1. Show the location and size of pipes and to locate and identify
     fittings, valves, fixtures and accessories that make up the
     piping system.

  2. Make an accurate estimate of costs.

  3. Illustrate the proper installation of the system.

Pipe working drawings must therefore be drawn to scale and include
all necessary details and dimensions.

Orthographic, oblique and axonometric are the types of projection
most commonly used in pipe drawings. A double line or a single
line representation of pipes, valves, fittings and pipe fixtures may
be used. Examples of a single line piping diagram for solar aided
systems (not drawn to scale) are shown in Figures 10.1 and 10.2.

Pipe sizes and lengths must be indicated in all working drawings.
Sizes are indicated by writing the nominal pipe bore near the pipe.
Location dimensions are also shown when necessary and are usually
laid off from centre lines. Dimensions of valves and fixtures are
standard and therefore seldom shown.

INSTALLATION SCHEDULING

The scheduling of the proper sequential steps in installing solar
heating and/or cooling systems during new home construction is
important to avoid unnecessarily difficult situations with
consequent increases in total costs.

The different steps in installing a typical liquid-heating solar
system are shown in Figures 10.3 (a through e). They are shown
properly synchronised with the usual operations undertaken in
constructing a new house. The building in this example contains an

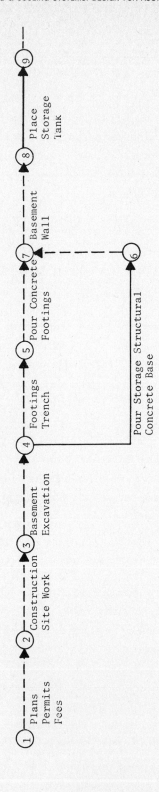

a. Storage concrete base at least 12.5 cm thick set on solid ground.

b. If storage tank is to be insulated, bottom insulation should be installed before placement.

Fig. 10.3 (a).   Placement of storage tank and foundation [1].

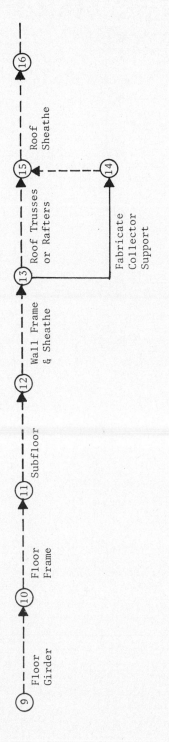

Fig. 10.3 (b). Collector support fabrication.

a. Collectors are usually mounted on plywood sheathing and then bolted to the rafters. In some cases, they can be bolted directly to the rafters.

b. Collectors mounted on flat roofs will require supports to attain proper orientation and tilt. Supports must be bolted to the rafters and enclosed to prevent wind drag and snow accumulation which can add additional loads on the roof.

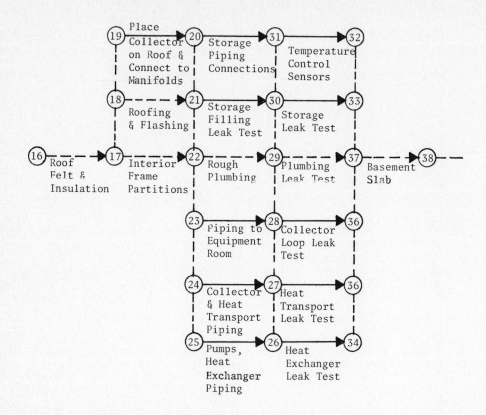

a. Collectors should be carefully inspected before
    installation for broken parts, dirty pipes and
    faulty fittings.

b. Side heat losses can be minimized by putting collectors
    side-by-side, otherwise insulation between collector
    modules is recommended.

Fig. 10.3 (c).   Installation and piping of collector
                 and auxiliary equipment [1].

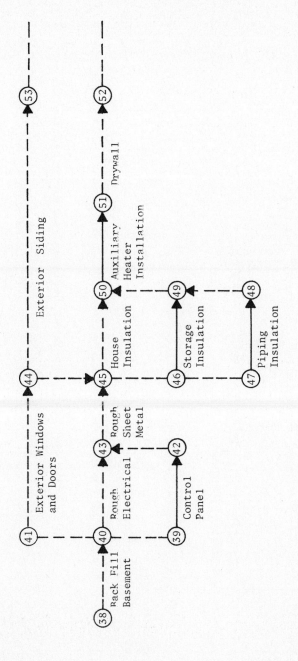

a. The storage tank, heat exchanger, expansion tank, pipes, as well as the valves should be properly insulated to minimize heat losses.

Fig. 10.3 (d). Insulation and auxiliary heater installation [1].

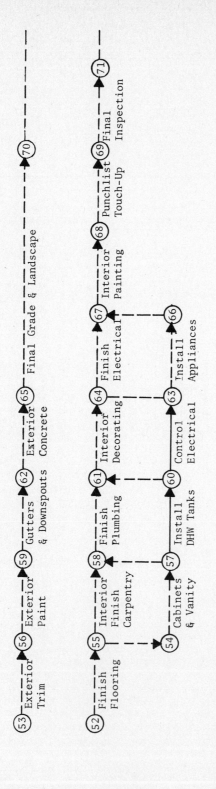

a. It is recommended that initial tests be made of the solar system and a final test and inspection be made after a short period of operation.

Fig. 10.3 (e). DWH tank, controls, tests and final inspection [1].

attached garage or storage room where the thermal storage unit will be located.

EQUIPMENT AND INSTALLATION SPECIFICATIONS

Equipment and installation specifications are an integral part of the design involving systems and are used to define the following:

1.   The work of the various trades.
2.   The kind and quality of materials involved
      in the construction and installation.
3.   The types of finish.
4.   The methods of installation and construction.

Specifications cover those points of installation and construction of systems that cannot be shown in the drawings.

SOLAR INSTALLATIONS

Thermosyphon Installation

Domestic hot water systems may be either of the simple thermosyphon or of a forced convection type. Typical collector areas range from 2 to 8 m^2.

Pipe sizing, slopes and insulation are critical factors in the performance of a solar system. The Australian Standards Association has recommended minimum pipe sizes and slopes for the simple thermosyphon system [2]. Figure 10.4 shows a typical thermosyphon system consisting of collectors, storage tank and pipes. Table 10.1 relates to Figure 10.4 and give minimum pipe sizes for a range of vertical and horizontal pipe lengths. Table 10.2 lists the parts of a domestic solar hot water system with recommended minimum insulation thicknesses.

Forced Convection Systems

When a thermosyphon is not practical, a forced convection system is

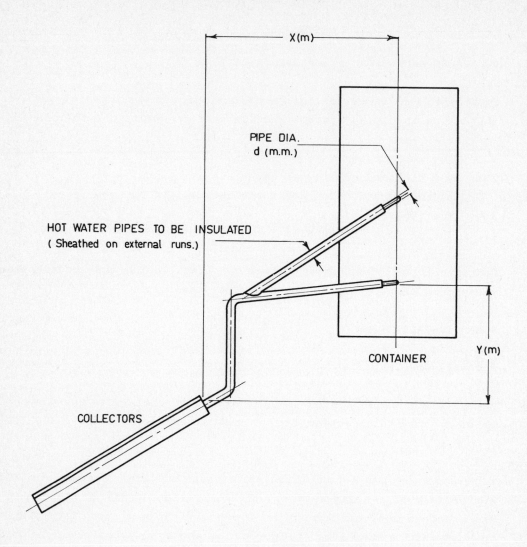

Notes: 1. Horizontal separation of pipes not shown.
        2. X and Y dimensions given in Tables 10.1 and 10.2
            are not pipe lengths but true horizontal and
            vertical distances, respectively.

        Fig. 10.4. Pipe sizes for thermosyphon system [2].

TABLE 10.1   Minimum Pipe Diameters* for a Thermosyphon System [2]

a. Collector Areas 2 - 4.5 m²

Y dimension (m)	X dimension (m)		
	1.5	3.0	7.0
	Pipe diameter d (mm)		
0.3	25	25	32
1.5	25	25	32
3.0	25	25	25

b. Collector Areas 4.5 - 8 m²

Y dimension (m)	X dimension (m)		
	1.5	3.0	7.0
	Pipe diameter d (mm)		
0.3	40	40	40
1.5	32	32	32
3.0	25	25	32

* Refer to Fig. 10.4

TABLE 10.2   Recommended Minimum Insulation of Hot Water Piping
[2]

Piping	Part to be Insulated	Thickness mm	
		Glass Wool	Flex. Pipe
Primary circuit, between the collector and the container	All piping	20	20
Secondary circuit	Flow and return	13	13
Draw-off (branch) hot water piping:			
(i) Branch from secondary circuit except droppers	Minimum 1 m from secondary circuit		
(ii) Branch from water heater except droppers	Minimum 1 m from the heater		
(iii) Floor mounted water heater	Minimum 1 m from the heater	13	13
(iv) Exposed to the weather (under freezing conditions)	All piping		
(v) Buried in the ground	All piping		
Cold water from feed tank to water heater	Minimum 1 m from the heater	13	13
Vent, exhaust and relief	Minimum 1 m from the heater	13	13
Vent, exhaust and relief drain pipes	Piping exposed to the weather (under freezing conditions)	13	13

used. Figure 10.5 shows such a system. This installation is typical of a mains pressure domestic hot water system. The detail in Figure 10.5 shows that the tank is a double walled, two compartment unit with the working fluid circulating in an outer annulus around the tank shell. With this arrangement the working fluid may be selected from a number of fluids such as:

1. Inhibited glycol with de-ionized or distilled water.
2. Silicon fluids.
3. Hydrocarbon oils (heat exchanger oils).

Larger Systems

When a large array of collectors is to be installed, then careful design of the pipework is essential to ensure that all the collectors can operate at their maximum performance. Investigations have shown [7] that poor flow distribution can exist if more than 16 risers are connected in parallel. Large arrays of collectors should be connected in parallel/series arrangements such as the ones shown in Figure 10.6. If there are a number of parallel/series arrangements in a particular loop, the fluid in the final array of collectors will be at a temperature that may be considerably higher than that of the first array. This will result in the final array of collectors operating at higher absorber plate temperatures, hence lower efficiencies. The flow rate through each array of collectors (Figure 10.6) should be adjusted so that the exit temperatures are equal for each array, thus a temperature sensor is needed at the exit of each array. An automatic air relief valve and a vacuum relief valve should also be installed at the highest point of the exit of each collector array.

Example of a Li-Br Solar Heating and Cooling System

Figure 10.2 shows a space heating and cooling system which is typical of larger installations and the equipment schedule is shown in Table 10.3. The system may be operated in four basic modes: daytime cooling, night cooling, day or night heating and auxiliary back-up.

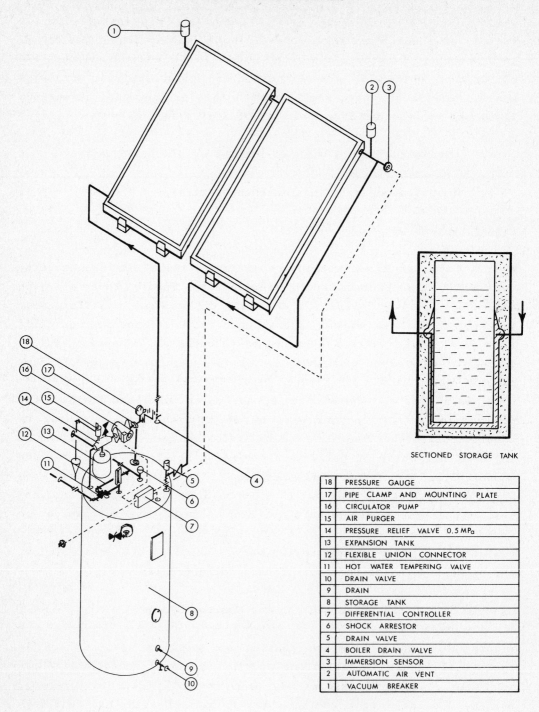

SECTIONED STORAGE TANK

18	PRESSURE GAUGE
17	PIPE CLAMP AND MOUNTING PLATE
16	CIRCULATOR PUMP
15	AIR PURGER
14	PRESSURE RELIEF VALVE 0.5 MPa
13	EXPANSION TANK
12	FLEXIBLE UNION CONNECTOR
11	HOT WATER TEMPERING VALVE
10	DRAIN VALVE
9	DRAIN
8	STORAGE TANK
7	DIFFERENTIAL CONTROLLER
6	SHOCK ARRESTOR
5	DRAIN VALVE
4	BOILER DRAIN VALVE
3	IMMERSION SENSOR
2	AUTOMATIC AIR VENT
1	VACUUM BREAKER

Fig. 10.5.   Forced convection system.

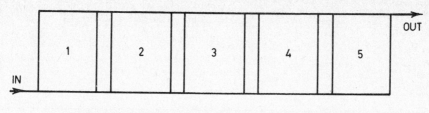

a. PARALLEL CIRCUIT CONNECTION (NOT RECOMMENDED)

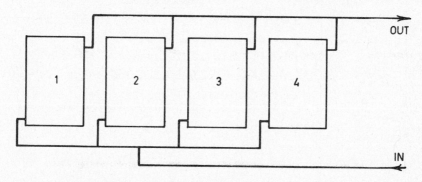

b. ALTERNATIVE PARALLEL CIRCUIT

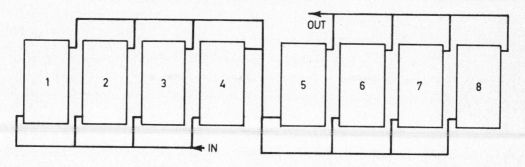

c. PARALLEL-SERIES CIRCUIT

Fig. 10.6.   Alternative circuit configurations.

TABLE 10.3   Equipment Schedule (Fig.9)

Item No.	Description
1.	Solar Collectors.
2.	Automatic air relief valve - Float vent.
3.	Flow control valve with indicator.
4.	Pump - Collector loop (size for 1 ℓ/min. per m^2 of collector area at system head).
5.	Air purger.
6.	Expansion tank with fill valve.
7.	Differential temperature controller.
8.	Level controller, float type.
9.	Hot storage tank, gal. steel 0.7 MPa W.P.
10.	Storage loop pump.
11.	Control Panel.
12.	Heat exchanger - Collector loop primary.
13.	Heat exchanger - Storage loop.
14.	Heat exchanger - Storage withdrawal loop.
15.	Pump - Storage withdrawal loop.
16.	3-Way solenoid valve.
17.	Pump - Chiller loop.
18.	Pump - Chilled water storage loop.
19.	Pump - System distribution.
20.	Bypass control valve.
21.	Proportional thermostat (conditioned space)
22.	Zone control valve (conditioned space).
23.	Valve - one way.
24.	Automatic timer control.
25.	Pump - Cooling tower loop.
26.	Pump - Back-up chiller loop.
27.	Chilled water storage tank.

Daytime cooling

Heat collected from the collectors (Figure 10.2) is transferred to
the absorption chiller via the collector loop primary heat exchanger
(12). Additional heat is transferred to hot storage (9) by the
storage loop heat exchanger (13), and the working fluid is pumped
back to the solar collectors (1).

The chiller unit delivers chilled water to the chiller water storage
tank (27) and rejects heat to the cooling tower. Chilled water is
then distributed from the storage tank (27) to the load zones upon
demand. A central control panel (11) positions the electrically
operated valves (16) to provide the correct flow path. Water flow
through the zone coil is varied by a modulating valve (22) in
response to a positional thermostat (21) mounted in the zone.
Proper sizing of the major system components will result in enough
energy stored daily in the hot water storage tank (9) to allow
several hours of night cooling.

Night cooling

The controls (11) automatically repositions the solenoid valve (16)
and activate the storage withdrawal pump (15). Stored heat is
transferred in the storage withdrawal heat exchanger (14) to the
chiller hot water loop to continue operating the chiller. When the
temperature in the hot storage tank falls below $75^{\circ}C$ the absorption
chiller is stopped. If the chilled water temperature in tank (27)
is below say $14^{\circ}C$ this chilled water is circulated to the zones by
pump (19).

Heating - day or night

Solar heat absorbed by the collectors is transferred to the hot
storage tank (9) via the storage loop heat exchanger (13) and pump
(4). Upon demand from any zone, the system distribution pump (19)
is activated by the control panel (11) and circulates heated water
through the distribution loop. Modulated flow through the zone
coil is set to meet the demand. Excess flow capacity of the

distribution pump (19) is recirculated around the pump by the bypass control valve (20).

Auxiliary back-up

Cooling back-up may be provided by a standard electrically-driven vapour compression unit which would deliver chilled water to storage (27).   Chilled   water   is   then   distributed   via   the   normal distribution loop.   This chiller would be activated any time the chilled water storage tank (27) rises above a preset temperature. Heating back-up could be provided by use of electric strip heaters mounted in the zone ductwork.   Alternatively, the vapour compresion unit (if a suitable selection is made) could be operated as a heat pump.   In this mode the condenser cooling water would be circulated to the hot storage tank (9).

Further energy recovery could be obtained during the summer months by using a further heat exchanger in the cooling tower water loop. This heat at say $38^{O}C$ would be satisfactory for preheating the incoming domestic or process water supply.

Insulation

Proper consideration must be given to the insulation of pipework, storage tanks and heat exchangers.   Chilled water pipework and storage tank require a vapour barrier on the external surface of the insulation.

Given the thermal conductivity of an insulating material, and the temperature of the fluid in a tank or pipe, the methods used in Chapter 2 can be used to calculate a thickness of the insulation material that will reduce heat losses or prevent unwanted heat gains.   However, calculation of the required pipe insulation is more complex and the manufacturer's catalogues should be consulted.

WARRANTIES ON SOLAR EQUIPMENT

Most equipment suppliers do not currently offer long-term warranties that are appropriate for solar energy applications. If a supplier provides any warranty at all, it is probably of the limited type, that is only certain features are covered for a limited period. For example, the two warranties given below are both limited, but provide quite different coverages:

Limited Warranty

This product is guaranteed against all defects in construction and against corrosion for a period of 5 years. Manufacturer will pay for all labour and parts costs to correct problems.

or

Limited Warranty

This product is guaranteed to be one of the finest solar systems ever manufactured. Manufacturer will pay for costs of parts to correct any problems.

Since the home owner will probably expect his solar heating systems to last at least 20 years before he replaces it, these short-term limited warranties are not enough.   Also, manufacturers usually will not guarantee the efficiency or heat delivery capability of solar equipment.   This is partly due to the lack of control the manufacturer has over the quality of the installation, the compatibility of his component with other components in the system, the variation in local climate and the practical difficulty of predicting and measuring the output of a specific system.   If the architect or builder is denied recourse to the manufacturer when the solar heat system malfunctions or does not perform as expected, law suits and, in some cases, bankruptcy can occur.

Many of the above difficulties are not insurmountable for the quantity producer of integrated units.   Suppliers of such packages for retrofit installation are able to rely on an extensive network of agencies and should be in a position to guarantee performance.

However, the same can not be said for one-off designs in which the collector and system are incorporated into the design of a building. The owner in this case has little redress in the event of poor performance and is in a position of carrying a substantial risk. Deficiencies in this case must be rectified component by component.

Limited warranty should be available on system components such as storage tanks, pumps, fans, coils and controls. These should be stipulated and performance of such components monitored as closely as instrumentation permits. Acceptance testing on these items is covered by appropriate standards (e.g. [3] for pumps) and furthermore appropriate dry runs can be organised for controls. Performance testing of systems and exposure testing of collectors were discussed in Chapter 5.

REFERENCES

1. Colorado State University Solar Energy Applications Laboratory, Solar Heating and Cooling of Residential Buildings, Sizing, Installation and Operation of Systems, U.S. Department of Commerce (1977).

2. Australian Standards Association, "Installation of Household-Type Solar Hot Water Systems", Standard No.2002-1979

3. I.S.O.2548, Centrifugal, Mixed Flow and Axial Pumps - Code for Acceptance Tests - Class C (1973).

4. Cooper, P.I., Symons, J.G. and Pott, P., "Thermal Performance Characteristics for Analysis, Design and Rating of Flat-Plate Solar Collectors", I.S.E.S. Solar Energy Congress, Los Angeles (1975).

CHAPTER 11

Performance Analysis

POWER GENERATION AND STORAGE

This chapter will consider some of the many factors that affect the economic performance of solar heating and cooling systems. Solar heating for swimming pools and domestic hot water (DHW) systems have proven that they are economic alternatives to conventional systems. In many regions properly designed solar home-heating systems are nearly competitive with conventional heating systems. Solar heating systems are also used in a few industrial applications. In these applications, special circumstances usually exist which give solar a competitive edge over the other alternatives. At the other extreme, there are still very few cases where solar air conditioning can compete economically with the conventional vapour-compression system.

It is the marginal system that is most likely to require a detailed economic analysis to determine its suitability for a particular application. Thus, this chapter will be concerned mainly with the economic factors that affect both solar home-heating systems and low-temperature commercial applications that could use solar. First, the economics of conventional electric heating systems will be discussed. Then, a life-cycle cost analysis will be performed for the solar-heated house studied in Chapters 2 and 6. Finally, the effect on the system economics of the assumptions made in the life-cycle analysis, as well as the effect tax of concessions and other factors, will be discussed.

ECONOMICS OF ELECTRIC POWER GENERATION

To predict the economic performance of solar heating systems, it is first necessary to understand the economics of conventional systems that compete against solar. In New South Wales, electrical heating systems now provide 75% of all home-heating requirements [1].

Consequently, in this chapter some of the special problems encountered by electricity commissions in providing electricity for home heating will be considered.

The relatively low population density in Australia and the abundance of coal and natural gas should keep the cost of electricity well below that of the rest of the world. This was the case during the 1970's when Australian electricity rates did not increase as much as the general inflation rate. However, with the increased demand for Australian coal and with the introduction in Australia of new energy hungry industries (such as aluminium smelters and uranium enrichment facilities) the cost of electricity increased much more rapidly between 1979 and 1983. Table 11.1 shows that the electricity rates charged to domestic users by the Sydney County Council increased by a factor of nearly three between 1979 and 1983.

TABLE 11.1   Domestic Electricity Rates Charged
by the Sydney County Council

YEARS	1979	1980	1981	1982	1983
price, c/kWh	2.65	3.13	3.84	7.15*	7.65*
% change	--	18	18	86	6.5

* Rate charged after the first 1000 kwh used in
that three-month period.

MANAGING THE UPS AND DOWNS OF ELECTRICAL USAGE

Regardless of hourly, daily and seasonal variations, there is a minimum level of demand that each generating authority prepares for on a year-round basis. This minimum demand level is called the "base load". Base-load plants tend to operate on a continuous basis with very little variation in output. They are usually the largest and most efficient in the system but, unfortunately, also the ones with the highest inertia (i.e. they respond slowly to changes in demand). The total system must also be capable of satisfying short-term changes in demand, so it usually has some units capable of fast response. Normally, either peaking or intermediate units are used for this purpose. Peaking units are used only when the demand is very high, which typically occurs just a few days each year. Since these units are idle most of the year,

economics dictate a sacrifice in energy efficiency in favour of low
capital cost.   One common approach is to use gas turbines as
peaking units since their capital costs are relatively low and they
can respond quickly to changes in demand.   However, they do have a
low efficiency and they only burn oil or gas (not coal).   Most
utilities in Australia and New Zealand have a few of these.

Between peaking units and base-load units there are a variety of
intermediate units, such as low efficiency fossil-fueled plants and
hydroelectric plants.   In N.S.W. and Victoria pumped-storage
hydroelectric plants in the Snowy Mountains Scheme are used to meet
much of the peak demand [2].   With pumped storage, water is pumped
from a low to a high reservoir at times of low system power demand
(when cheap electrical energy is available from base-load plants).
Afterwards, turbogenerators which can be quickly brought into
service use this stored water to meet peak demands.   In this way,
pumped-storage systems make effective use of available cheap power
from base-load plants.   However, some energy is wasted during this
process, since the overall efficiency of a pumped-storage system is
only between 65% and 75% (i.e., the electrical energy delivered from
the turbogenerator is only 65 to 75% of the electrical energy that
is received from the base load plants).

Since pumped-storage systems are extremely expensive and peaking
units are relatively inefficient, power authorities often offer
various incentives (e.g. off-peak rates) to try to smooth the demand
curve.   Another approach is to offer an "interruptable" power
service where the customer agrees to allow part of his electric
service to be cut off during peak demand times in exchange for lower
rates.   Both of these approaches work quite well in conjunction
with solar thermal-storage systems.   Since the energy is to be used
in the same form as it is stored, even if all of the energy stored
in the solar thermal-storage system comes from electricity, less
energy will be wasted than the 25 to 35% lost with pumped storage.
This will save energy and the cost of additional power plant
equipment which would otherwise be needed.

By encouraging the use of solar energy and energy conservation in
the home, it may be possible to eliminate or at least delay the need

for additional power plants.   For example, the Public Utilities Commission in California recently ordered its four main utility systems to start programs that encourage the use of solar water heaters and the weatherproofing of homes.   Over a three year period the utilities offered their customers interest-free loans for this purpose and about 200,000 homes were involved in the program.   As a result of these programs, the Pacific Gas and Electricity Company has been able to postpone plans for a large coal-powered station at the edge of San Francisco Bay, and the need that existed for a new $300M gasified-coal plant in the Mojave Desert is not as pressing as it was when the plant was first proposed.

Seasonal Variations

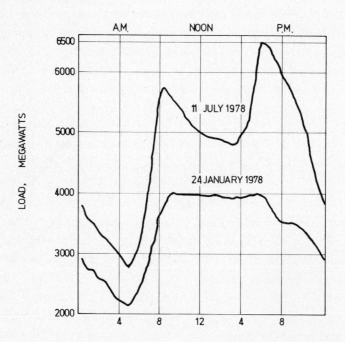

Fig. 11.1.   Summer and winter load curves
for days of maximum demand.

The demand for electrical energy varies throughout the year as well as throughout the day.   Figure 11.1 [3] shows that in New South Wales the demand is much greater during the winter months than during the summer.   Since the system is designed to have enough capacity to meet the maximum demand, much of this equipment stands

idle during the summer months when the maximum demand is seen to be
about one-third lower than the winter maximum. If it were possible
to reduce the demand for electricity used for heating during the
winter, less plant equipment would be needed and it would be used
more efficiently.   This would reduce both the capital and the
operating costs of the electricity generating plant.   The
introduction of solar home heating systems could have a significant
effect on the winter demand for electricity and might reduce both
the summer and winter differential as well as the maximum yearly
demand.

Daily Variations

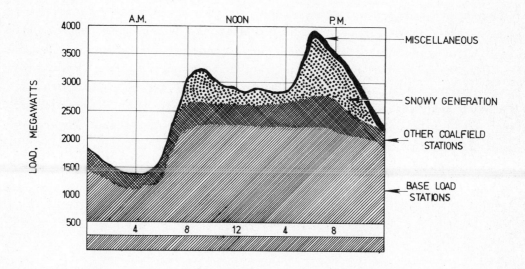

Fig. 11.2 Typical winter day load pattern.

In addition to seasonal variations in demand for electricity, the
demand also varies over a wide range during each 24 hour period.
The variation experienced in New South Wales during a typical day is

shown in Figure 11.2.   These variations in demand mean that the electrical generating plants cannot operate under steady-state conditions, but must cycle between low and high output conditions during each day.   Cycling of these units usually results in fuel being wasted.   It also makes it more difficult (and expensive) to control air pollution from the generating plants.   All of these factors increase the capital and the operating costs of the plant.

Role Played by Solar Home Heating

As indicated in the previous sections there are two ways solar home-heating systems might be utilized to reduce the daily variations in electrical demand during winter.   The first is the direct replacement of resistance-heating systems with solar-heating systems and the second is to encourage the installation of undersized solar-heating systems which would use electricity generated during off-peak hours to meet the excess heating demand. With this type of system, solar energy absorbed during the day would be used to satisify the heating requirements during the day and in the evening, and off-peak electricity would be used for heating between midnight and 4 a.m. (say) and to raise the temperature of the thermal-storage system high enough so that the demand for early morning heating could be provided from storage.

COST ANALYSIS OF SOLAR HEATING SYSTEMS

INTRODUCTION

The economics of a solar heating system are influenced by the application.   For example, the heating demand for a caravan park would have its peak demand during the summer tourist season while a home would require most of its heat during the cold winter months and a factory might have a relatively steady demand year-round. The cost analysis on the home heating system that follows illustrates most of the basic principles of this type of analysis and shows how the various factors interact.   In Appendix 1 an analysis of a caravan park located near Sydney illustrates the case where the maximum demand occurs in the summer.

In 1977 an American study of the cost of several solar home heating systems constructed within the previous four years showed that system-costs averaged about $270 (U.S.) [$A 300] per square metre of collector [4].   This study was conducted in Colorado on systems that were designed to supply about 75% of the home heating requirements.   A recent cursory study of similar equipment in Sydney [5] indicated that system costs here are probably not significantly less than this $A 300 figure.

The largest part of the initial cost of solar heating systems is for the collectors.   A survey of Australian suppliers [5] revealed that collector prices are not stable.   One imported collector that had been priced at approximately $A 120 per square metre in 1978 was quoted at nearly $A 200 per square metre at the time of the survey (5 years later).   The increase was largely due to inflation and the decrease in the value of the Australian dollar during that period. In another instance, a locally manufactured collector was priced 50% higher than another collector of similar design and performance. However, the survey did find that at least one manufacturer could supply collectors similar to the one used in the f-chart analysis in Chapter 8 at about $133 per square metre for a large quantity purchase.

The 160 m^2 Sydney home analysed in Chapter 6 required 30 square metres of this collector (which would cost around $A 4000) to supply 65% of the heat and hot water used during the year.   This cost is given in Table 11.2 along with estimates obtained for the other costs involved in assembling a space and hot water heating system for this home [5].

Table 11.2   Cost Estimates for a Solar Home Heating System
with 30 m^2 of Collector Area

Fixed costs
Heat Exchangers	500
Pumps	350
Controls and Sensors	300
Piping	1700
Other installation costs	500
subtotal	3350

Variable Costs
Collectors (30 m^2 at $133/m^2)	4000
Collector Supports ($20/m^2)	600
Storage Tank (2500 litres)	1150
Overhead and Profit (25% of total cost)	2900
subtotal	8650

Total $ 12 000

Consequently, in Sydney, a solar heating system for this house would cost around $12 000.   In addition to this initial cost, operating costs, interest charges, fuel costs (gas, oil or electricity), insurance, the scrap value of the system at the end of its life, taxes, etc., must be considered to determine the total cost of the system.   For the solar heating system to be economic, the total life-cycle cost must be about the same as or less than that of conventional systems.

LIFE-CYCLE COSTING

Real Cost of Interest on Loan

Life-cycle costing is an attempt to account for all present and future costs of the home heating system.   The real costs of any system involve cash flows at different times, so inflation and

interest charges are significant factors.   The true rate of
interest can be estimated by subtracting the inflation rate from the
interest rate charged by the lender.   For example, if the bank
charges 14% interest on the loan and the inflation rate is 11%, the
real interest is 14 - 11 = 3%.

System life

Another uncertainty in life-cycle analysis is the number of years
the system will last and the rate at which its performance will
degrade.   For a well designed system that has been properly
installed, the major components would probably have a life
expectancy of about 20 years.   Poor choice of materials, improper
design, inferior workmanship, or inadequate protection during
shipping, unloading and erection can reduce this life expectancy
significantly.   Failure to observe proper precautions for freeze
protection can also add to this uncertainty.

In Florida, solar domestic hot-water systems have been in operation
for up to 60 years.   An inspection (conducted before the 1978
freeze) of many systems that were over 6 years old showed that the
performance of most of these older systems had degraded
significantly [6].   As was explained in the first chapter, even
when the system is designed by an expert, a significant decrease in
performance with time is still likely.   However, for the
calculations that follow, it shall be assumed that the performance
of the system will not change during its 20-year life.

Maintenance and insurance

Since manufacturers, designers and builders at present have little
experience with solar heating or cooling systems, maintenance will
probably be a significant factor in the life-cost of these systems.
As described in Chapter 1, between July 1978 and April 1979 Argonne
National Laboratories has studied the performance of 66 solar
heating and cooling systems [8].   They found that 20% of the
systems froze, 28% experienced control-system failures, 21% had
interconnection failures and 25 of the 66 systems had a total of 47
collector problems.   In addition to these early failures, after the

systems have been in operation for many years, long-term failures will probably occur (such as those caused by corrosion, degradation of the solar liquid, fluid-induced scaling, thermally induced low-cycle fatigue, etc.).   Thus controls, hot-water tanks, fan motors and pumps would probably need to be replaced during the 20 year life of the system.   Also, glass breakage will probably occur. No firm figures are available, but an annual maintenance cost of about 1% of the total system cost might be reasonable for a well designed system.   For these calculations it will be assumed that the combined annual cost of maintenance and insurance is 1.5% of the initial cost of the total system.

Future energy costs

Another major variable, the future price of energy, also has a large effect on any cost analysis.   The rate of increase in energy costs depends not only on the general inflation rate, but also on the economic and political decisions of governments.   Since Australia has massive reserves of coal (for electricity) and natural gas, it is possible that some energy costs may increase at about the same rate as the general inflation.   The price of electricity, however, is largely determined by political forces and, as Table 11.1 shows, the electricity rates charged to the domestic consumer in Sydney have increased much faster than the inflation rate over the past few years.   In these calculations it will be assumed that electricity prices will increase 4% faster than the inflation rate.

Present-Worth Analysis

The "present worth" method is a technique used to analyse life-cycle costs.   With this method, all costs are put on a present-worth basis in terms of present dollars.   Basically, it compares the initial cost of the solar heating system with the current value of energy costs saved over the life of the system.

The present worth of all the energy that will be saved by the solar heating system (PWET) can be calculated from

$$PWET \quad = \quad (PEC)\,(L_d)\,(F)$$

where PEC is the present cost of energy, L_d is the total annual energy needed for heating, and F is the "fuel cycle factor". The fuel cycle factor (F) varies with the real cost of money (i_m), the real cost of fuel (i_f) and the expected operating life of the system (n). The equations that are used to determine the fuel cycle factor are given below for all three possible cases.

Case 1:

$$i_m \, > \, i_f$$

$$F \; = \; (1 + i_f)^{n-1}[(1 + i_e)^n - 1]/[i_e(1 + i_m)^n]$$

where

$$i_e \; = \; (i_m - i_f)/(1 + i_f)$$

Case 2:

$$i_M \; = \; i_f$$

$$F \; = \; n/(1 + i_m)$$

Case 3:

$$i_m \, < \, i_f$$

$$F \; = \; [(1 + i_E)^n - 1]/[i_E(1 + i_m)]$$

where

$$i_E \; = \; (I_f - i_m)/(1 + i_m)$$

The fuel cycle factor obtained for Case 3 has been plotted in Figure 11.3 for the special case of a real interest rate of 3%.

The present worth of the total energy saved by the solar heating system (PWES) is just the product of the solar fraction (f) and the present worth of the energy costs for the total time period (PWET).

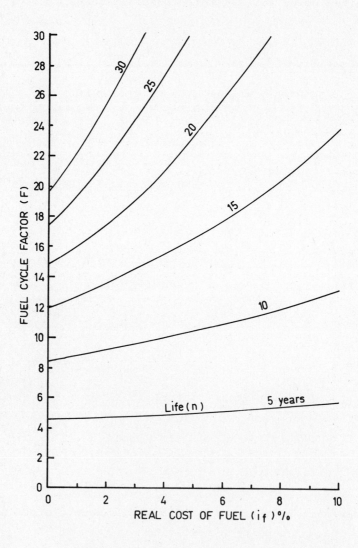

Fig. 11.3.   Variation of the fuel cycle factor with
             the real cost of fuel and the system life
             for a real interest rate of 3%.

That is,

$$PWES \quad = \quad (PWET)\,(f)$$

To illustrate this method, consider a solar home-heating system installed in Sydney that is expected to last 20 years (n = 20).   If over that period the average inflation rate is assumed to be 8%, the money interest rate 11% and energy costs to increase at a rate of 12%, then,

$$i_m \ = \ .03 \quad and \quad i_f \ = \ .04$$

and Case 3 applies since

$$i_m \ < \ i_f$$

So,

$$i_E \ = \ (0.04 - 0.03)/(1 + 0.03) \ = \ 0.0079$$

and

$$F \ = \ [(1 + 0.0097)^{20} - 1]/[0.0097(1 + .03)] \ = \ 21.31$$

Consequently, the present worth of what the energy would cost over the total 20-year period is 21.31 times the present cost of the conventional energy required for the first year of operation. Since the real interest rate (i_m) used in this example was 3%, the fuel cycle factor could have been read directly from Figure 11.3. From Figure 11.3, for i_m = 0.04 and n = 20, F is 21.3.

The f-chart analysis of this system (Chapter 6) indicated that 30 m^2 of collector at a tilt angle of 50° would have an annual solar fraction of 0.653.   How much can this system   cost and still compete with conventional systems?   The 160 square metre Sydney home mentioned earlier requires 63196 MJ of heating each year (Table 6.2), so the current annual cost for electric heating would be $1189 (using the current domestic Sydney County Council rate of $0.0765 per kWh).   As the solar heating system is expected to save 65.3% of this, the initial saving would be $776.42 per year. Using the lifetime factor of 21.31 calculated above, the present worth of the energy saved (PWES) is $18,694. This saving must be sufficient to cover the additional cost of the solar heating system as well as the difference in maintenance and in the insurance costs to justify its use.

Maintenance and insurance costs

If the fraction of the initial cost used each year for maintenance and insurance combined (i_c) is assumed to be 1.5% per year of the initial cost of the solar heating system (SCI), increasing along with general inflation, then the present worth of the combined maintenance and insurance expense (PWM) is determined from:

$$PWM = SCI \left[(1 + i_m)^n - 1\right] i_c / \left[i_m(1 + i_m)^n\right]$$

For this case, i_c = 0.015, so

$$PWM = 0.22 \ SCI$$

Thus, for this system and for the assumptions stated, installation of this solar heating system would be economically competitive if its initial cost (SCI) was,

$$SCI = PWES - PWM + [\text{Cost of electrical heating system}]$$

If the initial cost of an equivalent electrical resistance-heating system is assumed to be $2,000, then

$$SCI = 18,694 - 0.22 \ (SCI) + 2,000$$
$$SCI = \$ 16,918$$

This value is about 41% higher than the estimated $12,000 cost of the system. So, under the assumed conditions, solar heating is competitive with the electrical resistance heating system even today. The "pay-off" period can be calculated by adjusting the system life (n) in the fuel cycle factor equation until SCI is equal to the estimated system cost. The solar heating system would pay for itself in about 13 years.

These manual calculations are quite tedious, but the computer
program listed below can perform these calculations in less than one
second.   The program is in BASIC and is a simplified version of one
of the APPLE© programs on the SOLARAUST disk available from Pergamon
Press.   With this program it is a simple matter to optimise the
system and it is also relatively simple to perform a parametric
study of the various assumptions that were made during this analysis
(e.g. n, i_m, i_f, etc).

PRESENT-WORTH ANALYSIS

```
1   REM   THIS PROGRAM CALCULATES THE INITIAL COST OF LIQUID SOLAR
HEATING SYSTEM BY PRESENT WORTH ANALYSIS
10   PRINT "WHAT IS THE EXPECTED REAL COST OF MONEY OVER THIS
PERIOD?": INPUT RMC
20   PRINT "WHAT IS THE EXPECTED REAL COST OF ENERGY OVER THIS
PERIOD?": INPUT REC
30   PRINT "WHAT IS THE EXPECTED LIFE OF THE SYSTEM (IN YEARS)?":
INPUT N
40   PRINT "WHAT IS THE PRESENT COST OF ALTERNATE ENERGY (IN
$/KWH)?": INPUT PEC
50   PRINT "WHAT IS THE TOTAL ANNUAL ENERGY REQUIRED FOR THIS SITE
(IN MJ/YR)?": INPUT LT
60   PRINT "WHAT IS THE INSTALLED COST OF A CONVENTIONAL HEATING
SYSTEM (IN $)?": INPUT IAS
70   PRINT "WHAT FRACTION OF THE INITIAL COST OF THE SOLAR HEATING
SYSTEM SHOULD BE ALLOWED ANNUALLY FOR MAINTENANCE AND INSURANCE?":
INPUT MIC
80   PRINT "WHAT FRACTION OF THE TOTAL LOAD WILL BE PROVIDED BY SOLAR
(THE ANNUAL SOLAR FRACTION)?": INPUT ASF
90   PRINT
100 IF RMC > REC THEN 300
110 IF RMC = REC THEN 350
120 IE = (REC - RMC) / (1 + RMC)
130 FCF = ((1 + IE)^N - 1) / (IE * (1 + RMC))
140 PRINT
150 PRINT "THE FUEL CYCLE FACTOR IS    ";FCF
```

```
160 PWET = PEC * (LT / 3.6) * FCF
170 VESP = PWET * ASF
180 PRINT
190 PRINT "THE PRESENT WORTH OF THE ENERGY SAVED BY USING SOLAR IS $
";VESP
200 PWM = ((1 + RMC)∧N - 1) * MIC / (RMC * (1 + RMC)∧N)
210 SIC = (VESP + IAS) / (1 + PWM)
220 PRINT
230 PRINT "THE PRESENT WORTH ANALYSIS SHOWS THAT THE INITIAL COST OF
THIS SOLAR HEATING SYSTEM COULD BE AS HIGH AS $ ";SIC;
240 PRINT "  AND STILL BE ECONOMICALLY COMPETITIVE WITH EXISTING
SYSTEMS"
250 END
300 IE = (RMC - REC)/(1 + REC)
310 FCF = ((1 + IE)∧N - 1) * ((1 + REC)∧(N - 1)/(IE * (1 + RMC)∧N)
320 GOTO 140
350 FCF = N/(1 + RMC)
360 GOTO 140
```

OPTIMUM SYSTEM SIZE

In the analysis of a solar heating system a collector area of 30 m^2
was arbitrarily chosen to demonstrate the method.   Actually, the
collector area should have been selected to minimise the life-cycle
costs.   If the collector area is too small, the fixed costs in
Table 11.2 associated with piping and installation will not be
covered by the savings in fuel costs.   If the collector area used
is too large, much of the heat absorbed would not be needed when it
is available and would be wasted.   The optimum can be determined by
separating the total installed system cost into fixed costs and
those that vary with the collector area.   The only costs that vary
significantly with small changes in collector area are the variable
costs in Table 11.2 (i.e. the cost of the collectors themselves,
their supporting frames, the thermal-storage system, the overhead
and the profit).   Other costs, such as those for piping, ducts,
controls, pumps and blowers and for the architectural modifications
to the home, increase very little when the collector area is
increased.   Still other costs, such as the cost of the uncovered

part of the roof, may actually decrease when the collector area is increased. Several large programs have been written for mainframe computers which will determine the optimum sistem size. However, if the system is relatively small, such as a home heating systems, the optimum system size can be obtained manually by simply trying several values of collector area. This has been done for the system using the computer program above and it was found that the optimum system for this application will have a collector area around 30 m^2. This type of analysis produces results similar to those in Figure 11.4, which shows not only the variation of costs with collector size, but also the "breakeven" points.

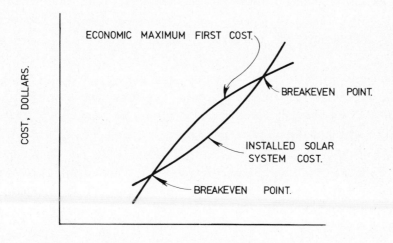

Fig. 11.4. Optimum system size.

Comments on the Cost Analysis

In this analysis a few unrealistic assumptions were made to simplify the calculations. Although these simplifications did not affect the primary objective of the book (which was to demonstrate the basic principles involved in the analysis of solar heating systems) they could have a large effect on any conclusions drawn from the results of the cost analysis. For example,

1.   It was assumed that the entire house was to be maintained at $21^{\circ}C$ for 24 hours every day.   An actual home would probably use zone heating and lower temperatures would be acceptable for at least part of the day.   Thus in Sydney, the total annual heat load for this type of house would be significantly less than the value used in this analysis.   If this annual heat load was reduced by 40%, the pay-off period would increase to 20 years.

2.   It was also assumed that the electricity prices would increase at a rate 4% greater than the general inflation rate for the next 20 years.   Since in Australia this electricity is generated from coal or natural gas which is readily available, it is possible that the price of electricity will increase at somewhat less than this rate. However, special factors, such as increased demand, could push electricity prices up even faster.

3.   The annual costs to run the pumps and fans in the solar heating system were neglected since these costs are dependent on the type of system used.   Typically, these costs might be between $10 and $70 per year at current electricity prices.

4.   Other factors that were neglected were the scrap value of the system after its 20 year life, and the value of the floor space occupied by the system (primarily the heat storage unit).

A similar analysis could have been done to compare the solar heating system with a reverse-cycle system.   Reverse-cycle systems require a larger initial capital investment than most other conventional systems, but they produce heat at about one-half the cost of the electric resistance heaters.   The reverse-cycle system would provide summer cooling as well as winter heating.   Consequently, to choose between a reverse-cycle and a solar-heating system, it would be necessary to balance cost against convenience.

Another possibility is to use off-peak electricity to provide at least part of the heat.   At the current off-peak rates of 3.04 cents per kWh the hot water and home heating costs could be reduced by up to one half.

Off-peak electricity might also be used to heat water in a large thermal storage tank.   Since this alternative would require a large and expensive storage system, it would probably not be competitive with electric resistance heating in Sydney.   However, by combining an undersized solar heating system (say 15 or 20 m^2 of collector area) with an off-peak electric system so that both use the same thermal storage system, the hybrid system may well be competitive with conventional heating systems.   A preliminary analysis indicates that 20 m^2 of collector area with a 1500 litre storage system would be adequate for the application considered in this chapter and would cost about $3000 less than the system considered. Although a detailed analysis has not been performed on this hybrid system, it appears that the pay-off period could be reduced to less than 10 years.

It would also have been possible to compare the solar heating system with a system that uses natural gas.   At present natural gas costs about the same as electricity for home heating but it is difficult to predict what natural gas will cost in a few years' time.

Regardless of what other factors are considered, the conclusion remains the same; that is, in Sydney, solar home-heating systems do not have a clear economic advantage over other heating systems at present.   However, the electricity generating authorities and the government might benefit economically and politically if they were to encourage solar heating in Australia.   Consequently, they may some day offer incentives to home owners installing solar home-heating systems.   Appendix 2 describes some of the tax incentives that have been offered in the U.S.A. since 1978 to encourage the use of solar energy.   It also describes the effect these incentives would have on the cost of the system analysed in this chapter.

Another possible incentive is for an electricity commission to offer off-peak electricity at the Tariff 1 rate of 3.04 cents per kWh for home heating to home-owners who install solar heating systems.   The effect of this would be to increase the present worth of energy saved (PWES) which would make the solar heating system much more attractive.

APPENDIX 1

ANALYSIS OF A SOLAR WATER HEATER FOR A CARAVAN PARK

As an example of the use of solar heating in a commercial
application consider the monthly variation in demand for a caravan
park located near Sydney.   This park will cater mainly to tourists
so its maximum demand will occur during the summer months and during
the school holidays.   The relevant design data for this site are
given in Figure 11.5 and Table 11.3.   The maximum monthly demand
occurs in both December and January and is 7500 litres per day of
hot water at 55°C.

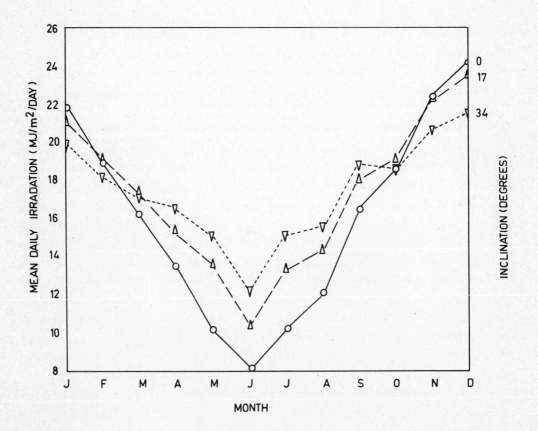

Fig. 11.5   Mean daily irradiation for an inclined surface in
Sydney.

Collector 2 in Table 4.3 has been selected for this application and 90 m^2 of this single-glazed, flat plate collector can fit on the existing roof.   During the peak-demand period (i.e. November through February) Figure 11.5 shows that a collector tilt angle of 0° and 17° would intercept the most solar radiation at this site. The 17° tilt angle was chosen since tilted collectors are easier to clean than horizontal collectors.

At 17° tilt the monthly average insolation on the collectors for that site is given in Column (c) and the heating demand (L) can be calculated from

$$L = Col.(b) \ (7500 \ litres/day) \ 4.18 \ kJ/(litre \ ^oC) \ [55 \ - \ Col.(a)]$$

TABLE 11.3   Performance of a Solar Water-Heating System in a Caravan Park

MONTH	(a) T (Mains) °C	(b) Demand / Demand in Dec.	(c) Insolation MJ/m^2 - Day	(d) L Demand MJ/Day	(e) L$_{SOLAR}$ MJ/Day
July	11.8	0.1	13.3	106	106
Aug.	13.1	0.1	14.2	106	106
Sept.	15.1	0.2	18.1	212	212
Oct.	17.6	0.4	19.1	424	424
Nov.	19.4	0.7	22.2	744	602
Dec.	21.1	1.0	23.5	1063	638
Jan.	22.0	1.0	21.2	1063	576
Feb.	21.9	0.8	19.0	850	516
Mar.	20.9	0.6	17.0	638	462
Apr.	18.3	0.4	15.4	424	418
May	15.1	0.4	13.2	424	358
June	12.8	0.1	10.4	106	106

Annual Totals

L (Demand) = 184,800 MJ/yr

L (Solar)   = 135,720 MJ/yr

For example, in December

$$L = 1.0 \ (7500) \ 4.18 \ [55 - 21.1]$$
$$= 1063 \ MJ/day$$

which is the value given for December in Column (d) of the table. An f-chart analysis shows that 90 square metres of collector number 2 will meet 100% or more of the demand between June and October and only a little over half of the demand for the peak months of December and January. This monthly solar contribution (L_{SOLAR}) is given in Column (e) of Table 11.3 and the annual solar fraction for the system is

$$f_{ANNUAL} = 135720/184800 = 74\%$$

So this system would provide 74% of the annual energy required for hot water at this caravan park. If the electricity rates for this park are 12.01 cents per kilowatt hour (the current commercial rate charged by the S.C.C.), the system would save $ 4,528 in its first year of operation. A relatively simple solar hot-water system (such as the one shown in Figure 10.1) is all that is needed for this application, so the total system cost could be less than $ 25,000 with a pay-off period of between 6 and 10 years. The economics of this system is likely to be significantly more favourable than for the home heating system since the peak demand occurs during the summer when the solar insolation is at its maximum. In addition, the cost of the system would be allowed as a tax deduction (i.e. a business expense). However, the interest rate charged for money lent to a business is usually much higher than that charged on a home mortgage, but again the business can deduct this expense from its taxes. So much more information is needed about the actual application before a more accurate estimate of its pay-off period could be obtained. Even though the economics of this type of system will vary significantly with the particular application, in this case the pay-off period could even be less than the 6 to 10 years indicated by the analysis.

APPENDIX 2

EFFECT OF TAX CONCESSIONS ON COST ANALYSIS

As John Yellott [8] pointed out, not only is "the power to tax the power to destroy", it is equally true that "the power to restrain taxation is the power to breathe life into a struggling enterprise". Many Australian states have considered offering tax incentives for solar installations.   For example, in May 1979 Premier Wran announced that New South Wales was considering rates rebates to encourage the use of solar energy.   Tax incentives of this type have been extremely effective in the United States.   It is still not known if tax rebates will be offered for solar heating systems in New South Wales or, if they are offered, what form they will take.   However, there are strong economic and social pressures on both State and Commonwealth Governments to offer something similar to some of the tax incentives that are presently offered in the U.S.A.   By the beginning of 1978, 30 different states in the U.S.A. had passed legislation to encourage the use of solar devices. These incentives usually took one of the following forms:

1.   Property tax (i.e. rate) exemptions or deductions.
2.   State income-tax credits or deductions.
3.   Sales tax exemptions.

In addition to the incentives offered by the states, in 1978, the U.S. Federal Government also introduced income tax rebates both for energy conservation measures and for solar-energy equipment used in the home.   The federal rebates were not deductions but tax refunds. In 1980 these rebates were increased and currently stand at 40% for up to $10,000 spent on solar equipment in any tax year.   For example, if the $12,000 system discussed above were installed in the U.S., and the costs spread over two tax years, the federal tax refund would been $4,800.   In addition, federal tax rebates of up to 15% were also given toward the cost of insulation, double glazing, caulking and weatherstripping of homes.

The incentives offered by individual American states are also significant. For example, if a California resident were to install this $12,000 solar heating system and spread the costs over two tax years, he could receive a tax rebate that would bring the total rebate to 55% of the actual cost (i.e. $6,600). In addition, he might also be eligible for low-interest government loans.

It is clear that tax incentives such as these could make solar home heating systems economically attractive for most home owners. Unfortunately, it appears that the main effect of the American incentives has been to make the solar industry more profitable.

NOMENCLATURE

Symbol	Meaning
f	fraction of total load provided by solar
F	fuel cycle factor
i_c	the fraction of the initial cost used each year for maintenance and insurance combined
i_m	real cost of money (i.e. interest rate minus the inflation rate)
i_f	real cost of energy (i.e. rate of energy minus the inflation rate)
L_d	energy needed without solar (kWh/yr)
n	total time period (years)
PEC	present energy cost ($/kWh)
PWET	present worth of what the energy will cost over the total time period ($)
PWES	present worth of total energy saved by using solar (heat load divided by the conversion efficiency)
PWM	the present worth of the combined maintenance and insurance expense ($)
SCI	the initial cost of the solar heating system ($)

REFERENCES

1. Ballinger, J. "Your House in the Sun", Univ. of N.S.W.
    Quarterly, No.13 (Dec. 1978) pp.12-13.

2. Thomas, M.H., "Energy Storage - Options and Economics",
    presented at The Institution of Engineers, Australia Meeting,
    Sydney Division (19 September 1979).

3. Electricity Commission of N.S.W., Annual Reports (1969-1978).

4. Colorado State University, "Solar Heating and Cooling of
    Residential Buildings: Design of Systems", U.S. Government
    Printing Office, Washington, D.C. (1977).

5. Survey conducted by the Heat-Transfer (Solar Design) class at
    The New South Wales Institute of Technology in June 1983.

6. Farber, E.A., University of Florida (private visit by E. Baker
    in 1978).

7. Chopra, P.S. "Reliability and Materials Performance of Solar
    Heating and Cooling Systems", Argonne National Labs. Report
    SOLAR/0906-79/70 (June 1979).

8. Yellott, .I., "Solar Energy Update", Heating/Piping/Air
    Conditioning (January 1979) pp.55-63.

Glossary

Absorber : That part of the solar collector that receives the solar radiation and converts it to a more useful form of thermal energy.

Absorptivity : The proportion of the radiation absorbed to the total incident radiation.

Air-Based Solar Heating System : A solar heating system in which air is used as the coolant in the solar collectors.

Air Stratification (Passive Design) : The warm air that collects at the ceiling of a room; can be removed by natural convection through vents (placed high on the walls) to assist in cooling the room.

Air Conditioning : To control the temperature, moisture content, movement and purity of air to meet the requirements of a conditioned space.

Ambient Temperature : Dry-bulb temperature of the outside air.

Building Structural (or Added) Mass : Mass within the building that provides natural thermal storage of sensible heat.

Capacitance Rate : The product of the mass flow rate and the specific heat of fluid flowing through the system.

Chiller : That part of the air conditioning system that cools the air.

Coefficient of Performance (COP) : The ratio of the cooling (or heating) capacity of the system to the energy put into the system (i.e. the energy paid for).

Collector : A component of the solar heating (or cooling) system which absorbs the solar radiation, converts it to useful thermal energy and transfers the thermal energy to a heat-transfer fluid.

Collector Efficiency : The ratio of the energy transferred to the solar collector fluid to the radiant energy incident on the collector during the same time period.

Collector Heat Removal Factor (F_R) : the ratio of the actual useful energy gain of a solar collector to the energy that would be gained if the entire plate were at the inlet temperature of the fluid.

Collector Overall Heat Loss Coefficient (U_L) : a parameter that represents the heat losses from the collector to the surroundings.

Conduction (Thermal) : Transfer of heat from a high-temperature region to a low-temperature region. The energy is transmitted by direct molecular interaction without appreciable displacement of the molecules.

Controls : System, based on devices such as thermostats for regulating the heat supply by the use of fans, pumps and dampers.

Convection : Transport of heat by fluid motion. In convection, temperature differences in the fluid cause density gradients which induce fluid motion from the warm region to the cool region. In forced convection, an external energy source such as a fan or a pump causes the fluid motion.

Cover Plate : A transparent material placed over a solar collector which transmits most of the solar radiation. Its primary purpose is to reduce the heat losses from the absorber to the surroundings.

Creep : Time-dependent deformation of a material under constant stress (or load).

Declination : The angle between the sun's rays and a plane through the equator.

Degree Day : One day with the mean of the daily maximum and minimum outside dry-bulb temperature one degree cooler (or warmer) than some specified base temperature.

Design Heating (or Cooling) Load : The maximum predicted heating (or cooling) requirements under design (i.e. near worst case) conditions.

Design Life : The period during which the system is expected to function without major maintenance or replacement.

Design Temperature : A chosen outside temperature that is close to the worst case condition. For example, 98% winter design temperature means that during an average winter the outside air temperature will not be lower than the chosen temperature for more than 2% of the 2160 hours in the three coldest months.

Design Temperature Difference : The difference between the inside and the outside design temperatures.

Diffuse Insolation : Solar radiation that has been scattered by atmospheric particles and gases and arrives at the earth's surface from all directions.

Direct Insolation : Solar radiation that comes in a straight line from the sun and has not been scattered.

Displacement Water Heater : A water heater in which cold water is fed into the container at or near the bottom, displacing but not mixing with the hot water as it is drawn off at, or near, the top.

Domestic Hot Water : hot water supply for conventional domestic purposes, such as washing and cooking.

Dry-Bulb Temperature : Temperature measured by an ordinary thermometer such as the mercury-in-glass thermometer or a thermocouple.

Emissivity : The proportion of radiation emitted by a body to that which would be emitted by a perfect emitter at the same temperature.

f-Chart : A graphical correlation that expresses the fraction of the monthly heating (or cooling) load that can be supplied by a solar energy as a function of the collector parameters and the local monthly average meteorological conditions.

Falling Level Water Heater : A water heater with a free water surface in which hot water is drawn off at or near the bottom, the level of the water falling as the hot water is drawn off.

Fenestration : The arrangement of windows in a building.

Flat-Plate Collector : A solar collector in which the absorber is a flat plate and which does not employ any external means of concentrating the sunlight.

Greenhouse Effect : Property of glass-enclosed structures to transmit short-wavelength (solar) radiation and to absorb the long-wavelength (infrared) radiation emitted from within the enclosure.

Heat Gain (Loss) : Heat gained in (or lost from) a building due to the sun, internal heat sources, infiltration and by conduction through the structure.

Hour Angle : 15° times the number of hours from solar noon

Humidity (Relative) : Approximately the ratio of the amount of water vapour in the air to the maximum amount of water vapour the air could hold at that dry-bulb temperature.

HVAC : Heating, ventilating and air conditioning.

Incidence Angle : the angle between the normal (or perpendicular) to a surface and the rays from the sun.

Inclination Angle : The angle between the absorbing surface and the horizontal.

Infiltration : Outside air that enters the house at an uncontrolled rate (through walls, cracks, around doors and windows, etc.) because of naturally occurring forces (e.g. wind).

Infrared Radiation : The long wavelength, radiant heat emitted by bodies at low temperatures (below 500 deg. C.)

Insolation : Radiation received by the sun.

Insulation : A material with a high thermal resistance which, when placed in the walls, ceilings or floors of a building will reduce the heat losses (or gains).

Latent Heat : Stored heat that is released when a vapour condenses or a liquid solidifies without a change in temperature.

Life-Cycle Costing : A comparison of future costs in terms of current money.

Liquid-Based Solar Heating System : A solar heating system which uses a liquid for the coolant in the collectors.

Long-Wavelength Radiation : (see infrared radiation)

Outgassing : Vapour release by materials, usually during exposure to high temperatures or low pressures.

Passive Solar Heating : Solar heating of a building by use of appropriate architectural details and construction techniques without the use of fans or pumps.

Phase-Change Thermal Storage : A thermal storage technique which uses the heat of fusion of a solid to store heat.

Pitting : Localized loss of material by erosion or electro-chemical decomposition.

Potable Water : Water suitable for drinking.

Preheater : A solar heater not containing a means of supplementary heating and installed to preheat the cold water supply prior to its entry into any other type of domestic water heater and which may be installed either with or without a bypass valved system.

Present Worth : A method of economic analysis in which all costs (present and future) are expressed in terms of current dollars.

Psychrometrics : Deals with the thermodynamic properties of moist air and the utilisation of those properties in the analysis of conditions and processes involving moist air.

Pyranometer : An instrument used to measure total (i.e. beam, diffuse and reflected) solar radiation.

Rock-Bed Thermal Storage : Large bins of sized rocks used to store sensible heat in air-based solar heating systems.

<u>Reflectivity</u> : The proportion of the radiation incident upon a surface that is reflected from that surface.

<u>Refrigeration</u> : Cooling or freezing fluids using mechanical or electrical devices.

<u>Retrofit</u> : Installation of a system into an existing structure that was not originally designed for that system.

<u>Selective Surface</u> : A surface with a high absorptivity for solar (short wavelength) radiation and a low emissivity for infrared (long wavelength) radiation.

<u>Sensible Heat</u> : Heat which is transferred that results in a change in temperature.

<u>Solar Degradation</u> : Deterioration of materials caused by exposure to sunlight.

<u>Solar Home-Heating System</u> : A system consisting of collectors to trap the solar radiation, fluid to transport the heat collected, a heat storage system and any necessary circulation and control systems.

<u>Sunset Hour Angle</u> : The hour angle that occurs at sunset.

<u>Temperature Relief Valve</u> : A temperature actuated valve which automatically discharges fluid at a specified set temperature. It is fitted to heaters to prevent the temperature in the container from exceeding a pre-determined temperature in the event that energy input controls fail to function.

<u>Thermosyphon</u> : Circulation established through difference in density between the hot and cooler portions of liquid.

<u>Thermosyphon Head</u> : Head due to the difference in density between hot and cooler portions of the liquid.

<u>Transmissivity</u> : The proportion of incident radiation that passes through the material unchanged except in direction.

<u>Wet-Bulb Temperature</u> : The air temperature measured with a thermometer that has a wetted wick around the end that senses the temperature. To obtain an accurate reading the air velocity around the wick should be between 5 and 10 metres per second.

Index